Lecture Notes in Computer Science   11843

Founding Editors

Gerhard Goos
Karlsruhe Institute of Technology, Karlsruhe, Germany
Juris Hartmanis
Cornell University, Ithaca, NY, USA

Editorial Board Members

Elisa Bertino
Purdue University, West Lafayette, IN, USA
Wen Gao
Peking University, Beijing, China
Bernhard Steffen
TU Dortmund University, Dortmund, Germany
Gerhard Woeginger ⓘ
RWTH Aachen, Aachen, Germany
Moti Yung
Columbia University, New York, NY, USA

Islem Rekik · Ehsan Adeli ·
Sang Hyun Park (Eds.)

Predictive Intelligence in Medicine

Second International Workshop, PRIME 2019
Held in Conjunction with MICCAI 2019
Shenzhen, China, October 13, 2019
Proceedings

 Springer

Editors
Islem Rekik 🆔
BASIRA
Istanbul Technical University
Istanbul, Turkey

Ehsan Adeli 🆔
Stanford University
Stanford, CA, USA

Sang Hyun Park 🆔
Daegu Gyeongbuk Institute
of Science and Technology
Daegu, Korea (Republic of)

ISSN 0302-9743   ISSN 1611-3349   (electronic)
Lecture Notes in Computer Science
ISBN 978-3-030-32280-9   ISBN 978-3-030-32281-6   (eBook)
https://doi.org/10.1007/978-3-030-32281-6

LNCS Sublibrary: SL6 – Image Processing, Computer Vision, Pattern Recognition, and Graphics

This Springer imprint is published by the registered company Springer Nature Switzerland AG
The registered company address is: Gewerbestrasse 11, 6330 Cham, Switzerland

Preface

It would constitute a stunning progress in medicine if, in a few years, we contribute to engineering a 'predictive intelligence' able to predict missing clinical data with high precision. Given the outburst of big and complex medical data with multiple modalities (e.g., structural magnetic resonance imaging (MRI) and resting function MRI (rsfMRI)) and multiple acquisition time-points (e.g., longitudinal data), more intelligent predictive models are needed to improve diagnosis of a wide spectrum of diseases and disorders while leveraging minimal medical data. Basically, predictive intelligence in medicine (PRIME workshop) primarily aims to diagnose in the earliest stage using minimal clinically non-invasive data. For instance, PRIME would constitute a breakthrough in early neurological disorder diagnosis as it would allow accurate early diagnosis using multimodal MRI data (e.g., diffusion and functional MRIs) and follow-up observations all predicted from only T1-weighted MRI acquired at baseline time-point.

Existing computer-aided diagnosis methods can be divided into two main categories: (1) analytical methods and (2) predictive methods. While analytical methods aim to efficiently analyze, represent, and interpret data (static or longitudinal), predictive methods leverage the data currently available to predict observations at later time-points (i.e., forecasting the future) or predicting observations at earlier time-points (i.e., predicting the past for missing data completion). For instance, a method that only focuses on classifying patients with mild cognitive impairment (MCI) and patients with Alzheimer's disease (AD) is an analytical method, while a method which predicts if a subject diagnosed with MCI will remain stable or convert to AD over time is a predictive method. Similar examples can be established for various neurodegenerative or neuropsychiatric disorders, degenerative arthritis, or in cancer studies, in which the disease or disorder develops over time.

Following the success of the first edition of PRIME-MICCAI in 2018, the second edition of PRIME 2019 aimed to drive the field of 'high-precision predictive medicine,' where late medical observations are predicted with high precision, while providing explanation via machine and deep learning, and statistically, mathematically, or physically-based models of healthy, disordered development, and aging. Despite the terrific progress that analytical methods have made in the last 20 years in medical image segmentation, registration, or other related applications, efficient predictive intelligent models and methods are somewhat lagging behind. As such predictive intelligence develops and improves (and this is likely to do so exponentially in the coming years), this will have far-reaching consequences for the development of new treatment procedures and novel technologies. These predictive models will begin to shed light on one of the most complex healthcare and medical challenges we have ever encountered, and, in doing so, change our basic understanding of who we are.

What Are the Key Challenges We Aim to Address?

The main aim of PRIME-MICCAI is to propel the advent of predictive models in a broad sense, with application to medical data. Particularly, the workshop accepted 8- to 12-page papers describing new cutting-edge predictive models and methods that solve challenging problems in the medical field. We hope that the PRIME workshop becomes a nest for high-precision predictive medicine – one that is set to transform multiple fields of healthcare technologies in unprecedented ways. Topics of interest for the workshop included but were not limited to predictive methods dedicated to the following topics:

- Modeling and predicting disease development or evolution from a limited number of observations
- Computer-aided prognostic methods (e.g., for brain diseases, prostate cancer, cervical cancer, dementia, acute disease, neurodevelopmental disorders)
- Forecasting disease or cancer progression over time
- Predicting low-dimensional data (e.g., behavioral scores, clinical outcome, age, gender)
- Predicting the evolution or development of high-dimensional data (e.g., shapes, graphs, images, patches, abstract features, learned features)
- Predicting high-resolution data from low-resolution data
- Prediction methods using 2D, 2D+t, 3D, 3D+t, ND, and ND+t data
- Predicting data of one image modality from a different modality (e.g., data synthesis)
- Predicting lesion evolution
- Predicting missing data (e.g., data imputation or data completion problems)
- Predicting clinical outcome from medical data (genomic, imaging data, etc)

Key Highlights

This workshop mediated ideas from both machine learning and mathematical, statistical, and physical modeling research directions in the hope of providing a deeper understanding of the foundations of predictive intelligence developed for medicine, as well as to where we currently stand and what we aspire to achieve through this field. PRIME-MICCAI 2019 featured a single-track workshop with keynote speakers with deep expertise in high-precision predictive medicine using machine learning and other modeling approaches which are believed to stand at opposing directions. Our workshop also included technical paper presentations, poster sessions, and demonstrations. Eventually, this helps steer a wide spectrum of MICCAI publications from being 'only analytical' to being 'jointly analytical and predictive.'

We received a total of 18 submissions. All papers underwent a rigorous double-blinded review process by at least 2 members (mostly 3 members) of the Program Committee composed of 26 well-known research experts in the field.

The selection of the papers was based on technical merit, significance of results, and relevance and clarity of presentation. Based on the reviewing scores and critiques, all PRIME 2019 submissions scored high by reviewers, i.e., all had an average score of at least above the acceptance threshold. Hence all 18 submissions were approved for publication in the present proceedings.

August 2019   Islem Rekik
                                                                                   Ehsan Adeli
                                                                                   Sang Hyun Park

Organization

Chairs

Islem Rekik	Istanbul Technical University, Turkey
Ehsan Adeli	Stanford University, USA
Sang Hyun Park	DGIST, South Korea

Program Committee

Amir Alansary	Imperial College London, UK
Changqing Zhang	Tianjin University, China
Dong Nie	University of North Carolina, USA
Duygu Sarikaya	University of Rennes 1, France
Gerard Sanroma	Pompeu Fabra University, Spain
Guorong Wu	University of North Carolina, USA
Heung-Il Suk	Korea University, South Korea
Ilwoo Lyu	Vanderbilt University, USA
Ipek Oguz	University of Pennsylvania, USA
Jaeil Kim	Kyungpook National University, South Korea
Le Lu	PAII Inc., USA
Lichi Zhang	Shanghai Jiao Tong University, China
Mara Valds Hernndez	University of Edinburgh, UK
Minjeong Kim	University of North Carolina at Greensboro, USA
Nesrine Bnouni	National Engineering School of Sousse (ENISo), Tunisia
Pew-Thian Yap	University of North Carolina, USA
Qian Wang	Shanghai Jiao Tong University, China
Qingyu Zhao	Stanford University, USA
Seyed Mostafa Kia	Donders Institute, The Netherlands
Stefanie Demirci	Technische Universität München, Germany
Sophia Bano	University College London, UK
Ulas Bagci	University of Central Florida, USA
Xiaohuan Cao	United Imaging Intelligence, China
Yu Zhang	Stanford University, USA
Yue Gao	Tsinghua University, China
Ziga Spiclin	University of Ljubljana, Slovenia

Contents

TADPOLE Challenge: Accurate Alzheimer's Disease Prediction Through Crowdsourced Forecasting of Future Data

Răzvan V. Marinescu[1,2]([✉]), Neil P. Oxtoby[2], Alexandra L. Young[2],
Esther E. Bron[3], Arthur W. Toga[4], Michael W. Weiner[5], Frederik Barkhof[3,6],
Nick C. Fox[7], Polina Golland[1], Stefan Klein[3], and Daniel C. Alexander[2]

[1] Computer Science and Artificial Intelligence Laboratory, MIT, Cambridge, USA
tadpole@cs.ucl.ac.uk
[2] Centre for Medical Image Computing, University College London, London, UK
[3] Biomedical Imaging Group Rotterdam, Erasmus MC, Rotterdam, The Netherlands
[4] Laboratory of NeuroImaging, University of Southern California, Los Angeles, USA
[5] Center for Imaging of Neurodegenerative Diseases, UCSF, San Francisco, USA
[6] Department of Radiology and Nuclear Medicine, VU Medical Centre,
Amsterdam, The Netherlands
[7] Dementia Research Centre, UCL Institute of Neurology, London, UK

Abstract. The Alzheimer's Disease Prediction Of Longitudinal Evolution (TADPOLE) Challenge compares the performance of algorithms at predicting the future evolution of individuals at risk of Alzheimer's disease. TADPOLE Challenge participants train their models and algorithms on historical data from the Alzheimer's Disease Neuroimaging Initiative (ADNI) study. Participants are then required to make forecasts of three key outcomes for ADNI-3 rollover participants: clinical diagnosis, Alzheimer's Disease Assessment Scale Cognitive Subdomain (ADAS-Cog 13), and total volume of the ventricles – which are then compared with future measurements. Strong points of the challenge are that the test data did not exist at the time of forecasting (it was acquired afterwards), and that it focuses on the challenging problem of cohort selection for clinical trials by identifying fast progressors. The submission phase of TADPOLE was open until 15 November 2017; since then data has been acquired until April 2019 from 219 subjects with 223 clinical visits and 150 Magnetic Resonance Imaging (MRI) scans, which was used for the evaluation of the participants' predictions. Thirty-three teams participated with a total of 92 submissions. No single submission was best at predicting all three outcomes. For diagnosis prediction, the best forecast (team *Frog*), which was based on gradient boosting, obtained a multiclass area under the receiver-operating curve (MAUC) of 0.931, while for ventricle prediction the best forecast (team *EMC1*), which was based on disease progression modelling and spline regression, obtained mean absolute error of 0.41% of total intracranial volume (ICV). For ADAS-Cog 13, no forecast was considerably better than the benchmark mixed effects model (*BenchmarkME*), provided to participants before the submission deadline. Further analysis can help understand which input

© Springer Nature Switzerland AG 2019
I. Rekik et al. (Eds.): PRIME 2019, LNCS 11843, pp. 1–10, 2019.
https://doi.org/10.1007/978-3-030-32281-6_1

features and algorithms are most suitable for Alzheimer's disease prediction and for aiding patient stratification in clinical trials. The submission system remains open via the website: https://tadpole.grand-challenge.org/.

Keywords: Alzheimer's disease · Future prediction · Community challenge

1 Introduction

Accurate prediction of the onset of Alzheimer's disease (AD) and its longitudinal progression is important for care planning and for patient selection in clinical trials. Early detection will be critical in the successful administration of disease modifying treatments during presymptomatic phases of the disease prior to widespread brain damage, i.e. when pathological amyloid and tau accumulate [1]. Moreover, accurate prediction of the evolution of subjects at risk of Alzheimer's disease will help to select homogeneous patient groups for clinical trials, thus reducing variability in outcome measures that can obscure positive effects on subgroups of patients who were at the right stage to benefit.

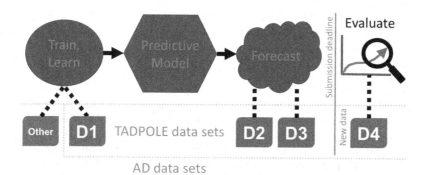

Fig. 1. TADPOLE Challenge design. Participants are required to train a predictive model on a training dataset (D1 and/or others) and make forecasts for different datasets (D2, D3) by the submission deadline. Evaluation will be performed on a test dataset (D4) that is acquired after the submission deadline.

Several approaches for predicting AD-related target variables (e.g. clinical diagnosis, cognitive/imaging biomarkers) have been proposed which leverage multimodal biomarker data available in AD. Traditional longitudinal approaches based on statistical regression model the relationship of the target variables with other known variables, such as clinical diagnosis [2], cognitive test scores [3], or time to conversion between diagnoses [4]. Another approach involves supervised machine learning techniques such as support vector machines, random forests, and artificial neural networks, which use pattern recognition to learn the relationship between the values of a set of predictors (biomarkers) and their labels (diagnoses). These approaches have been used to discriminate AD patients from cognitively normal individuals [5], and for discriminating at-risk individuals who convert to AD in a certain time frame from those who do not [6]. The emerging approach of disease progression modelling [7,8] aims to reconstruct biomarker trajectories or other disease signatures across the disease progression timeline, without relying on clinical diagnoses or estimates of time to symptom onset. Such models show promise for predicting AD biomarker progression at group and individual levels. However, previous evaluations within individual publications are not systematic and reliable because: (1) they use different data sets or subsets of the same dataset, different processing pipelines and different evaluation metrics and (2) over-training can occur due to heavy use of popular training datasets. Currently we lack a comprehensive comparison of the capabilities of these methods on standardised tasks relevant to real-world applications.

Community challenges have consistently proven effective in moving forward the state-of-the-art in technology to address specific data-analysis problems by providing platforms for unbiased comparative evaluation and incentives to maximise performance on key tasks. For Alzheimer's disease prediction in particular, previous challenges include the CADDementia challenge [9] which aimed to identify clinical diagnosis from MRI scans. A similar challenge, the "International challenge for automated prediction of MCI from MRI data" [10] asked participants to predict diagnosis and conversion status from extracted MRI features of subjects from the ADNI study [11]. Yet another challenge, The Alzheimer's Disease Big Data DREAM Challenge [12], asked participants to predict cognitive decline from genetic and MRI data. However, most of these challenges have not evaluated the ability of algorithms to predict clinical diagnosis and other biomarkers at future timepoints and largely used training data from a limited set of modalities. The one challenge that asked participants to estimate a biomarker at future timepoints (cognitive decline in one of the DREAM sub-challenges) used only genetic and cognitive data for training, and aimed to find genetic loci that could predict cognitive decline. Therefore, standardised evaluation of algorithms needs to be done on biomarker prediction at future timepoints, with the aim of improving clinical trials through enhanced patient stratification.

The Alzheimer's Disease Prediction Of Longitudinal Evolution (TADPOLE) Challenge aims to identify the data, features and approaches that are most predictive of future progression of subjects at risk of AD. The challenge focuses on forecasting the evolution of three key AD-related domains: clinical diagnosis, cognitive decline, and neurodegeneration (brain atrophy). In contrast to previous challenges, our challenge is designed to inform clinical trials through identification of patients most likely to benefit from an effective treatment, i.e., those at early stages of disease who are likely to progress over the short-to-medium term (defined as 1–5 years). Since the test data did not exist at the time of forecast submissions, the challenge provides a performance comparison substantially less susceptible to many forms of potential bias than previous studies and challenges. The design choices were published [13] before the test set was acquired and analysed. TADPOLE also goes beyond previous challenges by drawing on a vast set of multimodal measurements from ADNI which support prediction of AD progression.

This article presents the design of the TADPOLE Challenge and outlines preliminary results.

2   Competition Design

The aim of TADPOLE is to predict future outcome measurements of subjects at-risk of AD, enrolled in the ADNI study. A history of informative measurements from ADNI (imaging, psychology, demographics, genetics, etc.) from each individual is available to inform forecasts. TADPOLE participants were required to predict future measurements from these individuals and submit their predictions before a given submission deadline. Evaluation of these forecasts occurred post-deadline, after the measurements had been acquired. A diagram of the TADPOLE flow is shown in Fig. 1.

TADPOLE challenge participants were required to make month-by-month forecasts of three key biomarkers: (1) clinical diagnosis which is either cognitively normal (CN), mild cognitive impairment (MCI) or probable Alzheimer's disease (AD); (2) Alzheimer's Disease Assessment Scale Cognitive Subdomain (ADAS-Cog 13) score; and (3) ventricle volume (divided by intra-cranial volume). TADPOLE forecasts are required to be probabilistic and some evaluation metrics will account for forecast probabilities provided by participants.

3   ADNI Data Aggregation and Processing

TADPOLE Challenge organisers provided participants with a standard ADNI-derived dataset (available via the Laboratory Of NeuroImaging data archive at adni.loni.usc.edu) to train algorithms, removing the need for participants to pre-process the ADNI data or merge different spreadsheets. Software code used to generate the standard datasets is openly available on Github[1]. The challenge data includes: (1) CSF markers of amyloid-beta and tau deposition; (2) various imaging modalities such as magnetic resonance imaging (MRI), positron emission tomography (PET) using several tracers: FDG (hypometabolism), AV45 (amyloid), AV1451 (tau) as well as diffusion tensor imaging (DTI); (3) cognitive assessments such as ADAS-Cog 13 acquired in the presence of a clinical expert; (4) genetic information such as alipoprotein E4 (APOE4) status extracted from DNA samples; and (5) general demographic information such as age and gender. Extracted features from this data were merged into a final spreadsheet and made available online.

The imaging data was pre-processed with standard ADNI pipelines. For MRI scans, this included correction for gradient non-linearity, B1 non-uniformity

---

[1] https://github.com/noxtoby/TADPOLE.

correction and peak sharpening[2]. Meaningful regional features such as volume and cortical thickness were extracted using Freesurfer. Each PET image (FDG, AV45, AV1451), which consists of a series of dynamic frames, had its frames co-registered, averaged across the dynamic range, standardised with respect to the orientation and voxel size, and smoothed to produce a uniform resolution of 8 mm full-width/half-max (FWHM)[3]. Standardised uptake value ratio (SUVR) measures for relevant regions-of-interest were extracted after registering the PET images to corresponding MR images using SPM5. DTI scans were corrected for head motion and eddy-current distortion, skull-stripped, EPI-corrected, and finally aligned to the T1 scans. Diffusion tensor summary measures were estimated based on the Eve white-matter atlas.

3.1 TADPOLE Datasets

In order to evaluate the effect of different methodological choices, we prepared four "standard" data sets: the D1 standard training set contains longitudinal data from the entire ADNI history; the D2 longitudinal prediction set contains all available data from the ADNI rollover individuals, for whom challenge participants are asked to provide forecasts; the D3 cross-sectional prediction set contains a single (most recent) time point and a limited set of variables from each rollover individual – this represents the information typically available in a clinical trial; the D4 test set contains visits from ADNI rollover subjects after 1 Jan 2018, which contain at least one of the following: diagnostic status, ADAS score, or ventricle volume from MRI – this dataset did not exist at the time of submitting forecasts. Full demographics for D1–D4 are given in Table 1.

4 Submissions and Evaluation

The challenge had a total of 33 participating teams, who submitted a total of 58 forecasts from D2, 34 forecasts from D3, and 6 forecasts from custom prediction sets. Table 2 summarises the top-3 winner methods in terms of input features used, handling of missing data and predictive models: *Frog* used a gradient boosting method, which combined many weak predictors to build a strong predictor; *EMC1* derived a "disease state" variable aggregating multiple features together and then used an SVM and 2D splines for prediction, while VikingAI used a latent-time parametric model with subject- and feature-specific parameters – see [14] for full method details. We also describe three benchmark models which were provided to participants at the start of the challenge: (i) *BenchmarkLastVisit* uses the measurement at the last available visit, (ii) *BenchmarkME-APOE* uses a mixed effects model with APOE status as covariate and (iii) BenchmarkSVM uses an out-of-the-box support vector machine (SVM) and regressor for forecast.

---

[2] see http://adni.loni.usc.edu/methods/mri-analysis/mri-pre-processing.

[3] see http://adni.loni.usc.edu/methods/pet-analysis/pre-processing.

Table 1. Summary of TADPOLE datasets D1–D4. Each subject has been allocated to either Cognitively Normal, MCI or AD group based on diagnosis at the first available visit within each dataset.

Measure	D1	D2	D3	D4
Subjects	1667	896	896	219
Cognitively normal				
Number (% total)	508 (30%)	369 (41%)	299 (33%)	94 (42%)
Visits per subject	8.3 ± 4.5	8.5 ± 4.9	1.0 ± 0.0	1.0 ± 0.2
Age	74.3 ± 5.8	73.6 ± 5.7	72.3 ± 6.2	78.4 ± 7.0
Gender (% male)	48%	47%	43%	47%
MMSE	29.1 ± 1.1	29.0 ± 1.2	28.9 ± 1.4	29.1 ± 1.1
Converters (% total CN)	18 (3.5%)	9 (2.4%)	-	-
Mild cognitive impairment				
Number (% total)	841 (50.4%)	458 (51.1%)	269 (30.0%)	90 (41.1%)
Visits per subject	8.2 ± 3.7	9.1 ± 3.6	1.0 ± 0.0	1.1 ± 0.3
Age	73.0 ± 7.5	71.6 ± 7.2	71.9 ± 7.1	79.4 ± 7.0
Gender (% male)	59.3%	56.3%	58.0%	64.4%
MMSE	27.6 ± 1.8	28.0 ± 1.7	27.6 ± 2.2	28.1 ± 2.1
Converters (% total MCI)	117 (13.9%)	37 (8.1%)	-	9 (10.0%)
Alzheimer's disease				
Number (% total)	318 (19.1%)	69 (7.7%)	136 (15.2%)	29 (13.2%)
Visits per subject	4.9 ± 1.6	5.2 ± 2.6	1.0 ± 0.0	1.1 ± 0.3
Age	74.8 ± 7.7	75.1 ± 8.4	72.8 ± 7.1	82.2 ± 7.6
Gender (% male)	55.3%	68.1%	55.9%	51.7%
MMSE	23.3 ± 2.0	23.1 ± 2.0	20.5 ± 5.9	19.4 ± 7.2
Converters (% total AD)	-	-	-	9 (31.0%)

Table 2. Summary of benchmarks and top-3 methods used in the TADPOLE submissions. DPM – disease progression model. ($\dagger$) Aside from the three target biomarkers (*) Augmented features: e.g. min/max, trends, moments.

Submission	Extra[†] features	Nr. of features	Missing data imputation	Diagnosis prediction	ADAS/Vent. prediction
Frog	Most features	$70 + 420*$	None	Gradient boosting	Gradient boosting
EMC1-Std	MRI, ASL, cognitive	250	Nearest neighbour	DPM SVM 2D-spline	DPM 2D-spline
VikingAI-Sigmoid	MRI, cognitive, tau	10	None	DPM + ordered logit	DPM
BenchmarkLastVisit	-	3	None	Constant model	Constant model
BenchmarkME-APOE	APOE	4	None	Gaussian model	Linear mixed effects model
BenchmarkSVM	Age, APOE	6	Mean of previous values	SVM	Support vector regressor

For evaluation of clinical status predictions, we used similar metrics to those that proved effective in the CADDementia challenge [9]: (i) the multiclass area under the receiver operating curve (MAUC); and (ii) the overall balanced classification accuracy (BCA). For ADAS and ventricle volume, we used three metrics: (i) mean absolute error (MAE), (ii) weighted error score (WES) and (iii) coverage probability accuracy (CPA). BCA and MAE focus purely on prediction accuracy ignoring confidence, MAUC and WES include confidence, while CPA provides an assessment of the confidence interval only. Complete formulations for these can be found in Table 3, with detailed explanations in the TADPOLE design paper [13]. To compute an overall rank, we first calculated the sum of ranks from MAUC, ADAS MAE and Ventricle MAE for each submission, and the overall ranking was derived from these sums of ranks.

Table 3. TADPOLE performance metric formulas and definitions for the terms.

Formula	Definitions
$mAUC = \frac{2}{L(L-1)} \sum_{i=2}^{L} \sum_{j=1}^{i} \hat{A}(c_i, c_j)$	n_i, n_j – number of points from class i and j. S_{ij} – the sum of the ranks of the class i test points, after ranking all the class i and j data points in increasing likelihood of belonging to class i, L – number of data points
$BCA = \frac{1}{2L} \sum_{i=1}^{L} \left[\frac{TP}{TP+FN} + \frac{TN}{TN+FP} \right]$	TP_i, FP_i, TN_i, FN_i – the number of true positives, false positives, true negatives and false negatives for class i. L – number of data points
$MAE = \frac{1}{N} \sum_{i=1}^{N} \left| \tilde{M}_i - M_i \right|$	M_i is the actual value in individual i in future data. $\tilde{M}_i$ is the participant's best guess at M_i and N is the number of data points
$WES = \frac{\sum_{i=1}^{N} \tilde{C}_i \left| \tilde{M}_i - M_i \right|}{\sum_{i=1}^{N} \tilde{C}_i}$	M_i, $\tilde{M}_i$ and N defined as above. $\tilde{C}_i = (C_+ - C_-)^{-1}$, where $[C_-, C_+]$ is the 50% confidence interval
$CPA = |ACP - 0.5|$	Actual coverage probability (ACP) - the proportion of measurements that fall within the 50% confidence interval

5 Results

While full results can be found on the TADPOLE website [14], here we only include the top-3 winners. Table 4 compiles all metrics for top-3 TADPOLE forecasts from the D2 prediction set. The best overall performance was obtained by team *Frog*, with a clinical diagnosis MAUC of 0.931, ADAS MAE of 4.85 and Ventricle MAE of 0.45. Among the benchmark methods, *BenchmarkME-APOE* had the best overall rank of 18, obtaining an MAUC of 0.82, ADAS MAE of 4.75 and Ventricle MAE of 0.57. In terms of diagnosis predictions, *Frog* had an overall MAUC score of 0.931. For ADAS prediction, *BenchmarkME-APOE* had the best MAE of 4.75. For Ventricle prediction, *EMC1-Std* had the best MAE of 0.41 and WES of 0.29. In terms of the most accurate confidence interval estimates, *VikingAI* achieved the best CPA scores of 0.02 for ADAS and 0.2 for Ventricles.

Table 4. Ranked forecasting scores for benchmark models and top-3 TADPOLE submissions.

Submission	Overall	Diagnosis		ADAS			Ventricles (% ICV)		
	Rank	MAUC	BCA	MAE	WES	CPA	MAE	WES	CPA
Frog	1	0.931	0.849	4.85	4.74	0.44	0.45	0.33	0.47
EMC1-Std	2	0.898	0.811	6.05	5.40	0.45	0.41	0.29	0.43
VikingAI-Sigmoid	3	0.875	0.760	5.20	5.11	0.02	0.45	0.35	0.20
BenchmarkME-APOE	18	0.822	0.749	4.75	4.75	0.36	0.57	0.57	0.40
BenchmarkSVM	34	0.836	0.764	6.82	6.82	0.42	0.86	0.84	0.50
BenchmarkLastVisit	40	0.774	0.792	7.05	7.05	0.45	0.63	0.61	0.47

6   Discussion

In the current work we have outlined the design and key results of TADPOLE Challenge, which aims to identify algorithms and features that can best predict the evolution of Alzheimer's disease. Despite the small number of converters in the training set, the methods were able to accurately forecast the clinical diagnosis and ventricle volume, although they found it harder to forecast cognitive test scores. Compared to the benchmark models, the best submissions had considerably smaller errors that represented only a small fraction of the errors obtained by benchmark models (0.42 for clinical diagnosis MAUC and 0.71 for ventricle volume MAE). For clinical diagnosis, this suggests that more than half of the subjects originally misdiagnosed by the best benchmark model (*BenchmarkSVM*) are now correctly diagnosed with the new methods. Moreover, the results suggest that we do not have a clear winner on all categories. While team Frog had the best overall submission with the lowest sum of ranks, for each performance metric individually we had different winners.

Additional work currently in progress [14] suggests that consensus methods based on averaging predictions from all participants perform better than any single individual method. This demonstrates the power of TADPOLE in achieving state-of-the-art prediction accuracy through crowd-sourcing prediction models.

The TADPOLE Challenge and its preliminary results presented here are of importance for the design of future clinical trials and more generally may be applicable to a clinical setting. The best algorithms identified here could be used for subject selection or stratification in clinical trials, e.g. by enriching trial inclusion with fast progressors to increase the statistical power to detect treatment changes. Alternatively, a stratification could be implemented based on predicted "fast progressors" and "slow progressors" to reduce imbalances between arms. In order to make these models applicable to clinical settings, application in a clinical sample should be tested outside ADNI and further validation in a subject population with post-mortem confirmation would be desirable, as clinical diagnosis of probable AD only has moderate agreement with gold-standard neuropathological post-mortem diagnosis (70.9%–87.3% sensitivity and 44.3%–70.8% specificity, according to [15]). We hope such a validation will be possible in the future, with the advent of neuropathological confirmation in large, longitudinal, multimodal datasets such as ADNI.

In future work, we plan to analyse which features and methods were most useful for predicting AD progression, and assess if the results are sufficient to improve stratification for AD clinical trials. We also plan to evaluate the impact and interest of the first phase of TADPOLE within the community, to guide decisions on whether to organise further submission and evaluation phases.

Acknowledgement. TADPOLE Challenge has been organised by the European Progression Of Neurological Disease (EuroPOND) consortium, in collaboration with the ADNI. We thank all the participants and advisors, in particular Clifford R. Jack Jr. from Mayo Clinic, Rochester, United States and Bruno M. Jedynak from Portland State University, Portland, United States for useful input and feedback.

The organisers are extremely grateful to The Alzheimer's Association, The Alzheimer's Society and Alzheimer's Research UK for sponsoring the challenge by providing the prize fund and providing invaluable advice into its construction and organisation. Similarly, we thank the ADNI leadership and members of our advisory board and other members of the EuroPOND consortium for their valuable advice and support.

RVM was supported by the EPSRC Centre For Doctoral Training in Medical Imaging with grant EP/L016478/1 and by the Neuroimaging Analysis Center with grant NIH NIBIB NAC P41EB015902. NPO, FB, SK, and DCA are supported by EuroPOND, which is an EU Horizon 2020 project. ALY was supported by an EPSRC Doctoral Prize fellowship and by EPSRC grant EP/J020990/01. PG was supported by NIH grant NIBIB NAC P41EB015902 and by grant NINDS R01NS086905. DCA was supported by EPSRC grants J020990, M006093 and M020533. The UCL-affiliated researchers received support from the NIHR UCLH Biomedical Research Centre. Data collection and sharing for this project was funded by the Alzheimer's Disease Neuroimaging Initiative (ADNI) (National Institutes of Health Grant U01 AG024904) and DOD ADNI (Department of Defense award number W81XWH-12-2-0012). FB was supported by the NIHR UCLH Biomedical Research Centre and the AMYPAD project, which has received support from the EU-EFPIA Innovative Medicines Initiatives 2 Joint Undertaking (AMYPAD project, grant 115952). This project has received funding from the European Union Horizon 2020 research and innovation programme under grant agreement No. 666992.

References

1. Mehta, D., Jackson, R., Paul, G., Shi, J., Sabbagh, M.: Why do trials for Alzheimer's disease drugs keep failing? A discontinued drug perspective for 2010–2015. Expert Opin. Investig. Drugs **26**(6), 735 (2017)
2. Scahill, R.I., Schott, J.M., Stevens,. J.M., Rossor, M.N., Fox, N.C.: Mapping the evolution of regional atrophy in Alzheimer's disease: unbiased analysis of fluid-registered serial MRI. Proc. Natl. Acad. Sci. **99**(7), 4703–4707 (2002)
3. Yang, E., et al.: Quantifying the pathophysiological timeline of Alzheimer's disease. J. Alzheimer's Dis. **26**(4), 745–753 (2011)
4. Guerrero, R., et al.: Instantiated mixed effects modeling of Alzheimer's disease markers. NeuroImage **142**, 113–125 (2016)

5. Klöppel, S., et al.: Automatic classification of MR scans in Alzheimer's disease. Brain **131**(3), 681–689 (2008)
6. Young, J., et al.: Accurate multimodal probabilistic prediction of conversion to Alzheimer's disease in patients with mild cognitive impairment. NeuroImage Clin. **2**, 735–745 (2013)
7. Young, A.L., et al.: A data-driven model of biomarker changes in sporadic Alzheimer's disease. Brain **137**(9), 2564–2577 (2014)
8. Lorenzi, M., Filippone, M., Frisoni, G.B., Alexander, D.C., Ourselin, S., Alzheimer's Disease Neuroimaging Initiative, et al.: Probabilistic disease progression modeling to characterize diagnostic uncertainty: application to staging and prediction in Alzheimer's disease. NeuroImage **190**, 56–68 (2017)
9. Bron, E.E., et al.: Standardized evaluation of algorithms for computer-aided diagnosis of dementia based on structural MRI: the CADDementia challenge. NeuroImage **111**, 562–579 (2015)
10. Sarica, A., Cerasa, A., Quattrone, A., Calhoun, V.: Editorial on special issue: machine learning on MCI. J. Neurosci. Methods **302**, 1 (2018)
11. Weiner, M.W., et al.: Recent publications from the Alzheimer's disease neuroimaging initiative: reviewing progress toward improved AD clinical trials. Alzheimer's Dement. **13**(4), e1–e85 (2017)
12. Allen, G.I., et al.: Crowdsourced estimation of cognitive decline and resilience in Alzheimer's disease. Alzheimer's Dement. J. Alzheimer's Assoc. **12**(6), 645–653 (2016)
13. Marinescu, R.V., et al.: Tadpole challenge: prediction of longitudinal evolution in Alzheimer's disease. arXiv preprint arXiv:1805.03909 (2018)
14. https://tadpole.grand-challenge.org/Results/
15. Beach, T.G., Monsell, S.E., Phillips, L.E., Kukull, W.: Accuracy of the clinical diagnosis of alzheimer disease at national institute on aging alzheimer disease centers, 2005–2010. J. Neuropathol. Exp. Neurol. **71**(4), 266–273 (2012)

Inter-fractional Respiratory Motion Modelling from Abdominal Ultrasound: A Feasibility Study

Alina Giger[1,2](✉) ⓘ, Christoph Jud[1,2] ⓘ, Damien Nguyen[2,3] ⓘ,
Miriam Krieger[4,5] ⓘ, Ye Zhang[4] ⓘ, Antony J. Lomax[4,5], Oliver Bieri[2,3],
Rares Salomir[6], and Philippe C. Cattin[1,2] ⓘ

[1] Center for medical Image Analysis and Navigation,
University of Basel, Allschwil, Switzerland
{alina.giger,christoph.jud}@unibas.ch
[2] Department of Biomedical Engineering,
University of Basel, Allschwil, Switzerland
[3] Department of Radiology, Division of Radiological Physics,
University Hospital Basel, Basel, Switzerland
[4] Center for Proton Therapy, Paul Scherrer Institute (PSI),
Villigen PSI, Switzerland
[5] Department of Physics, ETH Zurich, Zurich, Switzerland
[6] Image Guided Interventions Laboratory,
University of Geneva, Geneva, Switzerland

Abstract. Motion management strategies are crucial for radiotherapy of mobile tumours in order to ensure proper target coverage, save organs at risk and prevent interplay effects. We present a feasibility study for an inter-fractional, patient-specific motion model targeted at active beam scanning proton therapy. The model is designed to predict dense lung motion information from 2D abdominal ultrasound images. In a pretreatment phase, simultaneous ultrasound and magnetic resonance imaging are used to build a regression model. During dose delivery, abdominal ultrasound imaging serves as a surrogate for lung motion prediction. We investigated the performance of the motion model on five volunteer datasets. In two cases, the ultrasound probe was replaced after the volunteer has stood up between two imaging sessions. The overall mean prediction error is 2.9 mm and 3.4 mm after repositioning and therefore within a clinically acceptable range. These results suggest that the ultrasound-based regression model is a promising approach for inter-fractional motion management in radiotherapy.

Keywords: Motion prediction · Ultrasound · 4D MRI · Radiotherapy

1 Introduction

Motion management is a key element in external beam radiotherapy of thoracic or abdominal tumours prone to respiratory movement. Pioneered in photon therapy, 4D treatment planning and motion monitoring techniques have gained in

© Springer Nature Switzerland AG 2019
I. Rekik et al. (Eds.): PRIME 2019, LNCS 11843, pp. 11–22, 2019.
https://doi.org/10.1007/978-3-030-32281-6_2

importance also in the field of particle treatments [13]. Due to higher dose conformity and the absence of radiation dose distal to the Bragg peak, proton therapy enables precise target treatment while spearing healthy tissue and organs at risk. However, in presence of organ motion, actively scanned proton beam therapies are hampered by interplay effects and inhomogeneous dose distributions [1] emphasising the need for sophisticated motion mitigation strategies, such as rescanning, gating or tracking [1,13]. In tracking, for example, the treatment beam is adapted to follow the tumour motion with the goal to ensure optimal target coverage. To do so, however, predictive methods and motion models are crucial in order to cope with respiratory motion variabilities and system latency.

In the field of radiotherapy, motion variabilities are classified into two categories: intra-fractional and inter-fractional motion variations [6]. Intra-fractional variations refer to motion variations between different respiratory cycles observed within a single treatment session; inter-fractional variations include anatomical and physiological differences between treatment sessions. Such motion variabilities should be considered for both treatment planning and dose delivery [5]. In this context, 4D imaging and motion modelling are widely discussed techniques. Motion models are necessary when direct imaging of the internal motion is not feasible. The idea is to estimate the motion of interest based on more readily available surrogate data. 4D imaging provides dense internal motion information and therefore constitutes an important element for respiratory motion modelling. While 4D imaging is traditionally performed with computed tomography (4D CT), respiratory-correlated magnetic resonance imaging (4D MRI) methods have increasingly been developed in the last decade due to their superior soft-tissue contrast and the lack of radiation dose [12].

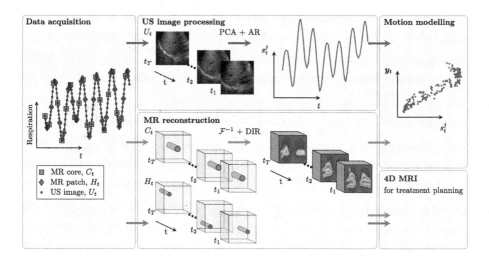

Fig. 1. Illustration of the pretreatment phase. See Sect. 2 and Sect. 3.1 for details.

In this work, we present an inter-fractional respiratory motion management pipeline for the lungs based on abdominal ultrasound (US) imaging as illustrated in Fig. 1. It involves hybrid US/MR imaging, principal component regression, and a novel 4D MRI technique [4]. The proposed approach follows a typical motion management scheme: In a pretreatment phase, simultaneous US and MR imaging acquisitions are performed and a motion model is computed. During treatment delivery, online US imaging is used to predict the respiratory motion for tumour tracking. We demonstrate the feasibility of our approach on five healthy volunteer datasets for two of which the US probe has been repositioned between motion modelling and prediction. Although not truly inter-fractional in the sense that there are days or weeks between two acquisitions, the presented data serve as preliminary data in this feasibility study.

US imaging has been proposed for image-guided interventions and radiotherapy before due to its advantages over other imaging modalities and surrogate signals [8]: it provides internal organ motion information at high temporal resolution, and therefore potentially detects phase shifts and organ drift [11], it is non-invasive and available during treatment delivery. However, as the lungs cannot be imaged directly, US guidance has mainly been applied for liver, heart or prostate. In [9], for example, an US-driven respiratory motion model for the liver has been presented. It requires precise co-registration of US and MR images in order to establish correspondence between tracked liver points. Indirect lung tumour tracking strategies based on 2D abdominal US have only been proposed recently [2,7]. Mostafaei et al. [7] combine US imaging and cone-beam CT (CBCT) in order to reduce the CBCT imaging frequency and therefore the imaging dose to the patient. However, the tumour motion is estimated in superior-inferior (SI) direction only. In [2] dense motion information was predicted based on an adversarial neural network. Although promising, it is not clear how this approach performs if the US imaging plane is shifted.

With this work we address the clinically relevant question of how the respiratory motion model performs in case of US probe repositioning between two imaging sessions. The novelty of our work does not primarily lie in the methodological components themselves but rather in their combination into a complete respiratory motion management pipeline. We combine US imaging with a recently presented 4D MRI technique and present first results in a feasibility study.

2   Background

Dense motion estimation is generally represented as a 3D deformation field which can be derived from any 4D imaging technique in combination with deformable image registration (DIR) methods. The 4D MRI sequence applied here uses 3D readouts and, unlike most other approaches, is a time-resolved imaging method [4]. As opposed to respiratory-correlated 4D MRI methods [12], it does not assume periodic respiration but provides continuous motion information. It is based on the assumption that the respiratory motion information is mapped mainly to the low-frequency k-space center. Following this rationale, circular patches at the k-space center $C_t \subset \mathbb{C}^3$ capture low-frequency image components with motion information while peripheral patches $H_t \subset \mathbb{C}^3$ account for image sharpness and structural details. Since these patches consist of a small portion of the k-space only, they can be acquired at a much higher temporal resolution as compared to the entire k-space. In Fig. 1, the 3D k-space is represented as a cube and the patches are illustrated as cylinders with the height pointing into the phase encoding direction.

Center and peripheral patches are acquired alternately and combined into patch pairs $P_t = \{C_t, H_t\}$. The center patches C_t are transformed to the spatial domain by applying the inverse Fourier transform $I_t = \mathcal{F}^{-1}(C_t)$. Then, a diffeomorphic registration method is applied to obtain the 3D deformation field between a reference image and I_t [10]. For further details, the reader is referred to [4]. In the following, we refer to the vectorised deformation field at time t as $\boldsymbol{y}_t \in \mathbb{R}^d$ with dimension d. Note that the peripheral patches H_t are not required for motion modelling but might be necessary for the treatment planning.

3   Method

3.1   Pretreatment Phase

Data Acquisition. Simultaneous US/MR acquisitions are performed in order to ensure temporal correspondence between the center patches C_t and the US images U_t as shown in Fig. 1. The US imaging plane is chosen such that parts of the liver and the diaphragm motion are clearly visible.

Image Processing and Reconstruction. Following the data acquisition, the 4D MRI is reconstructed and the motion vectors $\boldsymbol{y}_t$ are computed. Given 2D abdominal US images U_t, a low-dimensional respiratory motion surrogate is extracted using principal component analysis (PCA). By selecting only a small subset of principal components the model complexity is reduced. In order to cope with system latencies during dose delivery, it is important for the model to forecast the motion vectors into the future. Let $\boldsymbol{s}_t \in \mathbb{R}^k$ denote the standardised scores of the k most dominant principal components for image U_t. We apply an element-wise autoregressive (AR) model of order p for the time series $\{\boldsymbol{s}_t\}_{t=1}^T$:

$$s_t^j = \theta_0^j + \sum_{i=1}^p \theta_i^j s_{t-i}^j + \epsilon_t \quad \forall j \in \{1, \dots, k\}, \tag{1}$$

where s_t^j is the jth element of $\boldsymbol{s}_t$, $\boldsymbol{\theta}^j = \begin{bmatrix} \theta_0^j & \theta_1^j \dots \theta_p^j \end{bmatrix}^T$ denotes the model parameters, and ϵ_t is white noise. The parameters $\boldsymbol{\theta}^j$ are estimated using ordinary least squares. To predict the surrogate n steps ahead of time, the AR model in (1) is repeatedly applied.

Motion Modelling. In order for the motion model to capture non-linear relationships between the surrogates and the motion estimates, we formulate a cubic regression model. Let $\boldsymbol{x}_t \in \mathbb{R}^{3k+1}$ denote the input vector for the regression model which includes $\boldsymbol{s}_t$, its element-wise square and cube numbers, and a constant bias, i.e. $\boldsymbol{x}_t = \begin{bmatrix} 1 \ s_t^1 \ldots s_t^k \ (s_t^1)^2 \ \ldots \ (s_t^k)^2 \ (s_t^1)^3 \ \ldots \ (s_t^k)^3 \end{bmatrix}^T$. The motion model can thus be written as

$$\boldsymbol{y}_t = \boldsymbol{\beta}\boldsymbol{x}_t + \boldsymbol{\epsilon}_t, \tag{2}$$

with regression coefficients $\boldsymbol{\beta} \in \mathbb{R}^{d\times(3k+1)}$ and white noise $\boldsymbol{\epsilon}_t \in \mathbb{R}^d$. Given the pretreatment data $\{\boldsymbol{s}_t, \boldsymbol{y}_t\}_{t=1}^T$, the model parameters $\boldsymbol{\beta}$ are again approximated in the least-squares sense.

3.2   Online Motion Prediction

Having computed both the AR parameters in (1), and the regression coefficients in (2), the inference during dose delivery is straightforward and computationally efficient. However, since the motion modelling and treatment planning is performed several days or weeks prior to the dose delivery, the US probe has to be reattached to the patients' abdominal wall when they return for the treatment delivery. Although the location of the probe with respect to the patients chest can be marked by skin tattoos or similar approaches, it is hardly possible to recover the exact same imaging plane due to inter-fractional motions, anatomy changes, or different body positions with respect to the treatment couch [13]. The online US images can therefore not be projected onto the PCA basis directly, but a new principal component transformation has to be computed. We use the first minutes of US imaging after the patient has been setup for treatment as training data for recomputing a PCA basis. Since the first principal components capture the most dominant motion information and the scores $\boldsymbol{s}_t$ are standardised, we expect the signals to be comparable. Furthermore, the motion vectors $\boldsymbol{y}_t$ have to be warped in order to correspond to the present anatomy. This requires a 3D reference scan of the patients prior to treatment either using CT or MRI.

The surrogate signal $\boldsymbol{s}_t$ at time t is obtained by projecting the US image U_t onto the new PCA basis. Given the p latest surrogates $\{\boldsymbol{s}_{t-i}\}_{i=0}^{p-1}$, the signal $\boldsymbol{s}_{t+n}$ at time $t + n$ is approximated by applying the AR model n times. Finally, the motion estimate $\boldsymbol{y}_{t+n}$ is computed given Eq. (2) and warped in order to match the actual patient position.

4   Experiments and Results

Data Acquisition. The proposed motion management pipeline was tested on 5 healthy volunteers. The 4D MRI sequence [4] was acquired on a 1.5 T MR-scanner (MAGNETOM Aera, Siemens Healthineers, Erlangen, Germany) under free respiration and with the following parameters: TE $= 1.0$ ms, TR $= 2.5$ ms, flip angle $\alpha = 5°$, bandwidth 1560 Hz px^{-1}, isotropic pixel spacing 3.125 mm, image matrix $128 \times 128 \times 88$ and field of view $400 \times 400 \times 275\,\mathrm{mm}^3$ (in LR $\times$ SI $\times$ AP). The radius of C_t and H_t were set to 6 px and 5 px, respectively, resulting in 109 k-space points or 272.5 ms per center patch C_t, and 69 k-space points or 172.5 ms per peripheral patch H_t. The total acquisition time per subject was set to 11.1 min or $T = 1500$ center-peripheral patch pairs, P_t. For the reconstruction of the 4D MRI, a sliding organ mask was created semi-automatically [14].

US imaging was performed simultaneously at $f_{US} = 15\,\mathrm{Hz}$ on an Acuson clinical scanner (Antares, Siemens Healthineers, Mountain View, CA). A specifically developed MR-compatible US probe was attached to the patient's abdominal wall by means of a strap. The MRI and US systems were synchronised via optical triggers emitted by the MR scanner after every 6.675 s or 15 patch pairs P_t. The optical signal triggered the US device to record a video for a duration of 5 s. The time gap of 1.675 s was chosen to compensate for the US system latency while storing the video file. As a consequence, however, 4 patch pairs P_t per trigger interval are not usable due to missing US images. Despite this time gap, it sporadically happened that the trigger signal occurred before the preceding video file was stored resulting in an omission of the video just triggered. The time delay between the MR trigger and the start of the US video was negligible.

For subjects 4 and 5, the US probe was removed and reattached after they had been standing for several minutes. The US imaging plane was visually matched with the preceding imaging plane as good as possible. The MR images were aligned based on diffeomorphic image registration of two end-exhalation master volumes and inverse displacement field warping [3,10].

Table 1. Overview of the model settings and the respiratory motion characteristics for each subject s separately. The datasets with US probe repositioning are marked in grey.

s	model details			respiratory motion [mm]			
	AR model	motion model		training		test	
	train	train	test	μ_{95}	max	μ_{95}	max
1	600	640	200	13.31	40.92	15.65	40.96
2.1	600	470	200	5.68	29.81	4.65	29.96
2.2	600	470	200	5.68	29.81	12.83	45.76
3	600	530	200	4.35	19.80	4.95	19.03
4	600	660	200	6.92	24.49	7.32	23.92
5	600	630	200	5.63	25.74	4.87	16.34
4	–	–	690	6.92	24.49	6.99	27.48
5	–	–	830	5.63	25.74	6.43	25.31

Model Details. The first 8 US videos, corresponding to 200 images, were used to determine the AR parameters $\boldsymbol{\theta}$. The remaining data was split into a training and test set according to Table 1 in order to estimate $\boldsymbol{\beta}$ and validate the motion model performance, respectively. For each subject, the last 200 US/MR image pairs, or 133.5 s of data acquisition, were used for validation. For subject 2, however, a drastic change in respiratory motion characteristics was observed in the test set; the baseline motion more than doubled as compared to the training set. To take this observation into account, two test sets were created by dividing the last 267 s into equal parts. Below, the test set which includes deep respiratory motion, referred to as 2.2, is discussed separately. Table 1 shows the maximum and the baseline respiratory motion for each subject. The baseline motion μ_{95} is defined as the 95th percentile of the deformation field magnitude averaged over all time points.

For subjects 4 and 5, the parameters $\boldsymbol{\theta}$ and β were estimated based on the primary dataset. After the probe repositioning, the first 270 US images were used for recomputing the PCA basis. For all the experiments, the number of principal components was set to $k = 3$, and an AR model of order $p = 5$ was built. The surrogate $\boldsymbol{s}_t$ was predicted $n = 2$ steps, or $t_n = n/f_{US} = 133$ ms, into the future.

Validation. The predicted deformation field $\hat{\boldsymbol{y}}_t$ was compared to the reference $\boldsymbol{y}_t$. We define the prediction error as the magnitude of the deformation field difference for the masked region including the lungs as well as parts of the liver and the stomach. Figure 2 exemplarily illustrates the organ mask for volunteer 4 and 5 on a coronal slice of the master volume. In addition, the reference and the predicted deformation field magnitude, and the prediction error are shown. Highest motion magnitudes are observed in the region of the diaphragm. As expected, the prediction errors are higher for both volunteers after repositioning. It can be further observed that the motion model has a tendency to underestimate the respiratory motion. For volunteer 5, this becomes more evident when comparing the respiratory motion characteristics in Table 1 or Fig. 3: the respiratory motion has substantially increased after repositioning and therefore cannot be predicted precisely. Additionally, an organ drift of about 2 mm is observed in volunteer 5 after repositioning if all 830 test samples are considered which further decreases the prediction accuracy. The highest prediction errors are found at the lung boundaries.

Figure 3 shows the mean prediction error and the respiratory motion for the first 200 test samples. The shaded area marks the 5th and 95th percentile of the prediction error. The respiratory motion is defined as the 95th percentile of the reference deformation field magnitude. Since an end-exhalation master volume was used for registration, in general higher prediction errors are observed at end-inhalation. However, despite the decreased performance of the motion model after repositioning, the mean prediction error is substantially lower than the respiratory motion for most time points.

The box plots in Fig. 4 show the distributions of both the mean prediction error and the 95th percentile computed for each time point. Without US probe repositioning, the mean error is less or equal to 3 mm for all subjects except for volunteer 4 where it shows an outlier at 3.5 mm. The 95th percentile reaches a maximum value of 7.0 mm for volunteer 4 while 95% of the prediction errors for

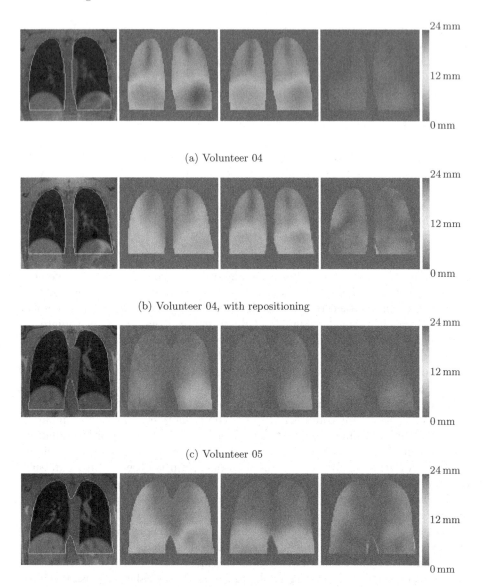

(a) Volunteer 04

(b) Volunteer 04, with repositioning

(c) Volunteer 05

(d) Volunteer 05, with repositioning

Fig. 2. Coronal cuts through sample end-inhalation volumes of volunteer 4 and 5 for both with and without repositioning. From left to right: master volume with the masked region marked in yellow, reference deformation field magnitude, predicted deformation field magnitude, and prediction error. (Color figure online)

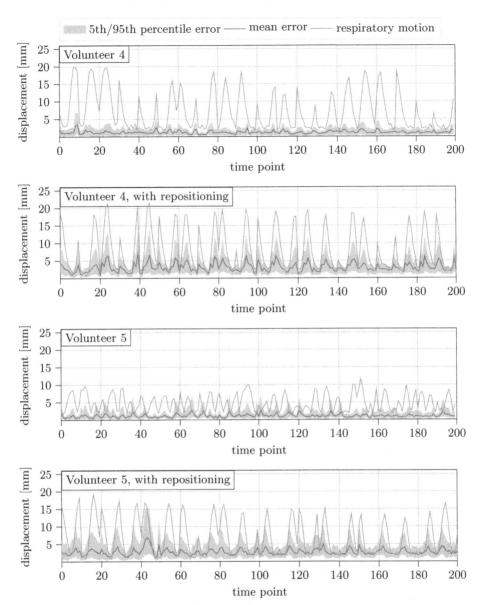

Fig. 3. Respiratory motion and prediction error over time for volunteer 4 and 5. For illustration purposes, only the first 200 test samples after repositioning are shown.

subjects 1, 2.1, and 3 are smaller than 6.0 mm, 5.4 mm, and 5.2 mm, respectively. The last column in Fig. 4 shows the results for the second test set of subject 2 where the respiratory motion was more pronounced as compared to the training

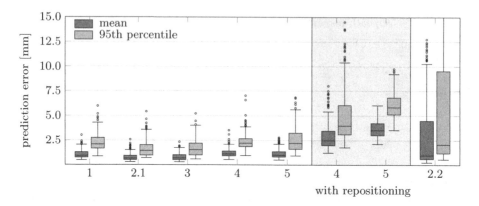

Fig. 4. Prediction error distribution for all volunteers; without (white background) and with (grey background) US probe repositioning. The whiskers of the box plots extend to the most extreme values within 1.5 times the interquartile range.

data. The maximum values for the mean prediction error and 95th percentile are 12.7 mm and 27.4 mm, respectively.

After US probe repositioning, the mean prediction error is below 8.0 mm and 6.0 mm for volunteers 4 and 5, respectively. There are, however, outliers of up to 14.5 mm for the 95th percentile of volunteer 4. By visual inspection of the prediction errors, it could be observed that these major discrepancies are located in the region of the stomach at the organ mask boundaries. In summary, the overall mean prediction error is 2.9 mm and 3.4 mm for volunteers 4 and 5, respectively.

5 Discussion and Conclusion

In this feasibility study we examined the performance of abdominal US surrogate signals in combination with a novel 4D MRI technique for lung motion estimation. The model predicts dense motion information 133 ms into the future which allows for system latency compensation. The obtained results are similar in terms of accuracy to those presented in previous studies [2, 9]. However, we additionally present preliminary findings for inter-fractional motion modelling which involves a repositioning of the US probe. Although the accuracy decreased when compared to intra-fractional modelling, overall mean prediction errors of 2.9 mm and 3.4 mm demonstrate that the proposed US surrogate signal is suitable even if the imaging plane is not identical for two fractions.

The presented results should, however, be treated with caution as the repositioning of the US probe has only been tested on two healthy volunteers and the time interval between the two measurements was in the range of minutes rather than days or weeks. Also, there exists no real ground-truth data for the respiratory motion. The reference deformation field might itself be corrupted due to registration errors. An additional error source is introduced with the alignment of the MR volumes between the two imaging sessions. Since this transformation was computed based on two exhalation master volumes, it might not be accurate for other respiratory states. Moreover, it was observed that the motion model does not generalise well if the respiration characteristics vary substantially as it was the case for subject 2. Although this limitation is inherent to the problem formulation and occurs in most motion models, it demands further investigations and characterisation. Also, further work is necessary to investigate the effect of dense motion predictions on treatment plan adaptations and dose distribution in proton therapy.

Acknowledgement. We thank Pauline Guillemin from the University of Geneva, Switzerland, for her indispensable support with the data acquisition. This work was supported by the Swiss National Science Foundation, SNSF (320030_163330/1) and the GPU Grant Program of NVIDIA (Nvidia Corporation, Santa Clara, California, USA).

References

1. Bert, C., Durante, M.: Motion in radiotherapy: particle therapy. Phys. Med. Biol. **56**(16), R113 (2011)
2. Giger, A., et al.: Respiratory motion modelling using cGANs. In: Frangi, A.F., Schnabel, J.A., Davatzikos, C., Alberola-López, C., Fichtinger, G. (eds.) MICCAI 2018. LNCS, vol. 11073, pp. 81–88. Springer, Cham (2018). https://doi.org/10.1007/978-3-030-00937-3_10
3. Jud, C., Giger, A., Sandkühler, R., Cattin, P.C.: A localized statistical motion model as a reproducing kernel for non-rigid image registration. In: Descoteaux, M., Maier-Hein, L., Franz, A., Jannin, P., Collins, D.L., Duchesne, S. (eds.) MICCAI 2017. LNCS, vol. 10434, pp. 261–269. Springer, Cham (2017). https://doi.org/10.1007/978-3-319-66185-8_30
4. Jud, C., Nguyen, D., Sandkühler, R., Giger, A., Bieri, O., Cattin, P.C.: Motion aware MR imaging via spatial core correspondence. In: Frangi, A.F., Schnabel, J.A., Davatzikos, C., Alberola-López, C., Fichtinger, G. (eds.) MICCAI 2018. LNCS, vol. 11070, pp. 198–205. Springer, Cham (2018). https://doi.org/10.1007/978-3-030-00928-1_23
5. Krieger, M., Giger, A., Weber, D., Lomax, A., Zhang, Y.: PV-0422 consequences of respiratory motion variability in lung 4DMRI datasets. Radiother. Oncol. **133**, S219–S220 (2019)
6. McClelland, J.R., Hawkes, D.J., Schaeffter, T., King, A.P.: Respiratory motion models: a review. Med. Image Anal. **17**(1), 19–42 (2013)
7. Mostafaei, F., et al.: Feasibility of real-time lung tumor motion monitoring using intrafractional ultrasound and kV cone beam projection images. Med. Phys. **45**(10), 4619–4626 (2018)
8. O'Shea, T., Bamber, J., Fontanarosa, D., van der Meer, S., Verhaegen, F., Harris, E.: Review of ultrasound image guidance in external beam radiotherapy part II: intra-fraction motion management and novel applications. Phys. Med. Biol. **61**(8), R90 (2016)
9. Preiswerk, F., et al.: Model-guided respiratory organ motion prediction of the liver from 2D ultrasound. Med. Image Anal. **18**(5), 740–751 (2014)
10. Sandkühler, R., Jud, C., Andermatt, S., Cattin, P.C.: Airlab: autograd image registration laboratory. arXiv preprint arXiv:1806.09907 (2018)

11. von Siebenthal, M., Szekely, G., Gamper, U., Boesiger, P., Lomax, A., Cattin, P.: 4D MR imaging of respiratory organ motion and its variability. Phys. Med. Biol. **52**(6), 1547 (2007)
12. Stemkens, B., Paulson, E.S., Tijssen, R.H.: Nuts and bolts of 4D-MRI for radiotherapy. Phys. Med. Biol. **63**(21), 21TR01 (2018)
13. Trnková, P., et al.: Clinical implementations of 4D pencil beam scanned particle therapy: report on the 4D treatment planning workshop 2016 and 2017. Physica Med. **54**, 121–130 (2018)
14. Vezhnevets, V., Konouchine, V.: GrowCut: interactive multi-label ND image segmentation by cellular automata. In: Proceddings of Graphicon, vol. 1, pp. 150–156. Citeseer (2005)

Adaptive Neuro-Fuzzy Inference System-Based Chaotic Swarm Intelligence Hybrid Model for Recognition of Mild Cognitive Impairment from Resting-State fMRI

Ahmed M. Anter[1,2]([✉]) [iD] and Zhiguo Zhang[1] [iD]

[1] School of Biomedical Engineering, Health Science Center, Shenzhen University,
Shenzhen 518060, China
anter@szu.edu.cn
[2] Faculty of Computers and Information, Beni-Suef University,
Benisuef 62511, Egypt
ahmed_anter@fcis.bsu.edu.eg

Abstract. Individuals diagnosed with mild cognitive impairment (MCI) are at high risk of transition to Alzheimer's disease (AD). Resting-state functional magnetic resonance imaging (rs-fMRI) is a promising neuroimaging technique for identifying patients with MCI. In this paper, a new hybrid model based on Chaotic Binary Grey Wolf Optimization Algorithm (CBGWO) and Adaptive Neuro-fuzzy Inference System (ANFIS) is proposed; namely (CBGWO-ANFIS) to diagnose the MCI. The proposed model is applied on real dataset recorded by ourselves and the process of diagnosis is comprised of five main phases. Firstly, the fMRI data are preprocessed by sequence of steps to enhance data quality. Secondly, features are extracted by localizing 160 regions of interests (ROIs) from the whole-brain by overlapping the Dosenbach mask, and then fractional amplitude of low-frequency fluctuation (fALFF) of the signals inside ROIs is estimated and used to represent local features. Thirdly, feature selection based non-linear GWO, chaotic map and naive Bayes (NB) are used to determine the significant ROIs. The NB criterion is used as a part of the kernel function in the GWO. CBGWO attempts to reduce the whole feature set without loss of significant information to the prediction process. Chebyshev map is used to estimate and tune GWO parameters. Fourthly, an ANFIS method is utilized to diagnose MCI. Fifthly, the performance is evaluated using different statistical measures and compared with different met-heuristic algorithms. The overall results indicate that the proposed model shows better performance, lower error, higher convergence speed and shorter execution time with accuracy reached to 86%.

Keywords: rs-fMRI · MCI · Optimization · Swarm intelligence · ANFS · Chaos theory

© Springer Nature Switzerland AG 2019
I. Rekik et al. (Eds.): PRIME 2019, LNCS 11843, pp. 23–33, 2019.
https://doi.org/10.1007/978-3-030-32281-6_3

1 Introduction

Alzheimer's disease (AD) is the most common cause of dementia and it is a heavy burden to the patients and the society. Individuals with MCI are considered to be at a much higher risk of developing AD, with a yearly conversion rate of 15–20%. Therefore, early detection of MCI is important for possible delay of the progression of MCI to moderate and severe stages. Recently, using rs-fMRI for detection of MCI has gained popularity because of the simple experimental procedure and readily available data analytics tools [1]. Particularly, many studies have investigated neuroimaging and machine learning (ML) techniques for the early detection of MCI and conversion from MCI to AD on the benchmark datasets [2–6].

Generally, ML has been successful implemented in many application areas. Despite the efficacy of ML on the diagnosis of AD and MCI, there are still various problems, such as falling in local optimum, not being able to converge to optimum solution, sensitivity to noise, and data uncertainty. Among various difficulties in detection of MCI from rs-fMRI, the most challenging one should be the curse of dimensionality [7]. Discovering valuable hidden knowledge in large scale data has become a pressing matter. Generally, there are two famous approaches for feature selection: filter approach is based on criterion to score features and wrapper approach contains ML techniques as a part of the evaluation process. These methods suffer from recession and stuck in local minima as well as computationally expensive. In order to further improve the effect of feature selection, the global optimization algorithm is needed [8].

In this paper, a new hybrid intelligent optimization model based on binary version of grey wolf optimizer (GWO), Chebyshev chaotic map, and ANFIS is proposed for rs-fMRI-based MCI diagnosis. The proposed model comprised of five main phases. (1) Preprocessing: fMRI data are sliced, aligned, filtered, normalized, and smoothed to enhance data quality. (2) Feature extraction: 160 regions of interest (ROIs) are localized from the whole-brain fMRI data by overlapping the Dosenbach mask. Afterwards, fractional amplitude of low-frequency fluctuation (fALFF) of the signals inside ROIs are estimated and used to represent local features. (3) Feature selection strategy: non-linear grey wolf optimizer (GWO) based on Chebyshev chaotic maps and naive Bayes (CBGWO) is used to determine the significant ROIs based power spectrum features. GWO is a metaheuristic optimization technique inspired by the hunting strategy and behavior of the grey wolves. The integration between these techniques is able to improve the classification accuracy of the MCI dementia. (4) MCI classification: hybrid fuzzy logic and artificial neural network (ANN) called (ANFIS) method is utilized to diagnose MCI based on the features selected from CBGWO. (5) Performance evaluation: different evaluation criteria are used for feature selection and classification phases. This model utilizes the strong ability of the global optimization of the GWO which avoids the sensitivity of local optimization. In order to minimize error, maximize classification accuracy, reduce erroneous clinical diagnostic. The following research problems for features selection and optimization will be addressed in this study.

- How to maintain several agents involved in the architecture?
- How to plan an integrated solution for a given MCI problem?
- How to avoid over-fitting using the integrated optimized model?
- How can get best local optima in the model without stuck in local minima?
- How to converge fast and decreasing time consuming?
- Is the global optimization algorithms need to parameters tuning?

The remaining of this paper is organized as follows. Section 2 presents the background knowledge and the newly-developed CBGWO model. Section 3 introduces the fMRI dataset and the pipeline of CBGWO-ANFIS model for MCI diagnosis. Experimental results are presented in Sect. 4. Finally, Sect. 5 concludes this paper and presents the future work.

2 Methodology

2.1 Grey Wolf Optimization Algorithm (GWO)

GWO is a meta-heuristic technique, which was presented in 2014 [9]. GWO is inspired from the hunting task and behavior of the grey wolves. Grey wolves are structured in a group from five to eleven members. They regulate the set into four kinds of wolves' categories (alpha (α), beta (β), omega (ω), delta (δ)) through hunting process, to remain the discipline in a group. The mathematical model that represents the strategy process of encircling the prey through the hunting is expressed with details in [9]:

$$X_{t+1} = X_{p,t} - A * D \tag{1}$$

$$d = |CX_{p,t} - X_t| \tag{2}$$

$$A = 2br_1 - b \tag{3}$$

$$C = 2r_2 \tag{4}$$

where X_{t+1} is the position of the wolf, $X_{p,t}$ is the position of the prey, d is a variation vector, r_1 and r_2 are random parameters between [0 1]. A and C are the main two coefficient vectors used to control the exploration and exploitation of GWO, and b is a linearly reducing vector from 2 to 0 over course of iterations, indicated in Eq. (5).

$$b = 2 - 2(\frac{t}{T_{max}}) \tag{5}$$

where t is the current iteration, and T_{max} is the maximum number of iterations.

2.2   The New Hybrid Kernel-Based CBGWO Model

The main challenging task in GWO is how to balance between exploration and exploitation with random distribution behavior. For nonlinear features, exploration and exploitation are in conflict and there is no clear boundary between them. For this reason, GWO suffers from low convergence and falling in local minima. Therefore, Chebyshev chaotic map (CCM) [13] is used for updating GWO position instead of the random behavior. In the new CBGWO method, CCM and naïve Bayes (NB) are used to tackle the problems of GWO.

GWO Dynamic Parameters Tuning: The main problem in GWO algorithm is a random distribution in dimensional space, which leads to the unbalance between the two main factors exploration and exploitation due to the stochastic nature of GWO and it can fall into the local minima. In this paper CCM is used to improve the performance of CBGWO in terms of convergence speed and local minima avoidance. In the standard GWO, two essential parameters (A, C) are used to control the behavior of the algorithm. CCM is utilized to tune these parameters by replacing the random generators variables as shown in Eqs. (6–8).

$$b_{CCM} = \{(a_i - t)\left(\frac{a_f - a_i}{Max_{iter}}\right).CCM(t) \tag{6}$$

$$A_{ChoasVec} = 2b.CCM(t) - b \tag{7}$$

$$C_{ChoasVec} = 2.CCM(t) \tag{8}$$

where a_i, a_f are the elementary and latest values adjusted as 2, 0 respectively, t is the current iteration and Max_{iter} are the maximal number of iterations. b is linearly decreased from 2 to 0 based on CCM. CCM(t) is Chebyshev choatic map which is non-invertible, non-repetition and ergodicity of chaos. CCM is sequence of one dimensional choatic vector at iteration t and can be expressed by the following equation.

$$X_{i+1} = cos\left(tcos^{-1}(w_i)\right) \tag{9}$$

where i is an iteration number and the chaotic time series $x_i \in [0, 1]$ are obtained.

NB-Based Kernel/Fitness Function: To select the discriminative features from the search process, non-parametric kernel density estimation is utilized as a learning tool using naïve Bayes for evaluating the predictive potential of the features from four different perspectives; size of the feature subset, classification accuracy, stability, and generalization of the feature set [10].

The following objective/fitness function is used to calculate the fitness value for each wolf in the search space.

$$Fitness = (1 - \omega) * (1 - Acc) + \omega * \frac{As}{(Al - 1)} \tag{10}$$

where ω is an initialized factor equal to 0.01, Acc is the accuracy of corrected data classified to the predicted label obtained from NB classifier, As is the summation of candidate attributes, and Al is the attribute length. The goodness of each wolf position in the feature space is assessed by this objective function. The main criteria are satisfied by minimizing the objective function.

3 Application on fMRI-Based MCI Recognition

3.1 Participants and Data Acquisition

All Individuals were diagnosed and scanned by expert doctors with more than 20 years' experience in dementia and radiology. The fMRI dataset was from the First Affiliated Hospital of Guangxi University of Chinese Medicine using Simens Magnetom Verio 3.0T°. The repetition time (TR) $= 2000\,\mathrm{ms}$ and echo time (TE) $= 30\,\mathrm{ms}$, field of view (FOV) $= 240\,\mathrm{mm} \times 240\,\mathrm{mm}$, imaging matrix $= 64 \times 64$ slice, slice thickness $= 5\,\mathrm{mm}$, slices $= 31$, flip angle $= 90°$, total volumes $= 180$. In this study a total 127 subjects, 62 MCI (18 females, age: 66.66 ± 7.44; 44 males, age: 63.73 ± 5.94) and 65 NC (24 females, age: 66 ± 4.99; 41 males, age 64.12 ± 6.059), were recruited. All patients have written informed consent and informed about experimental content.

3.2 Pipeline and Phases of CBGWO-ANFIS

The aims of this paper are to develop new hybrid intelligence optimization model for MCI diagnosis from resting-state fMRI data. Figure 1 shows the schematic pipeline of the proposed CBGWO-ANFIS optimization model for MCI diagnosis. Basically, it is constructed using five main phases as follow:

1. **Pre-processing phase.** The fMRI preprocessing procedures include: removal of the first 5 volumes, slice timing, and head motion correction. The data are sliced, realigned, normalized, and smoothed to reduce the effects of bad normalization and to increase the signal-to-noise-ratio. Default mask (brainmask.nii) was used with a threshold of 50%. Several nuisance signals are regressed out from the data by utilizing the Friston 24-parameters model. Also, other resources of spurious variance are removed from the data through linear regression.
2. **Feature extraction phase.** 160 ROIs are localized using the Dosenbach pattern. fMRI signals inside the ROIs are presented as time-series. The fractional-ALFF (f-ALFF) method is used to transform time-series data into frequency domain and the power spectrum was obtained. The band-pass filter is performed to reduce the influences of low-frequency drift and high-frequency physiological noise. Then, principal component analysis (PCA) method is used to decrease the high correlation of this non-linear data.
3. **Feature selection phase.** CBGWO based on naive Bayes is used to determine the significant attributes in order to obtain the optimal features that represent the best ROIs for MCI. CCM is used to estimate and tune GWO parameters which dominate the exploration and exploitation rate.

4. **Classification phase.** The ANFIS method is used to provide an effective learning solution for MCI diagnosis [11].
5. **Performance evaluation phase.** The performance of the proposed model is evaluated using different statistical measures for feature selection and classification: Mean fitness (μ_f), Best fitness (BF), Standard deviation (Std), and Average Attributes Selected (AAS) for feature selection process. Accuracy (ACC), Precision, Recall, F1-score, Informedness ($Info$), Markedness ($Mark$), Matthews Correlation Coefficient (MCC), and kappa for classification process [12].

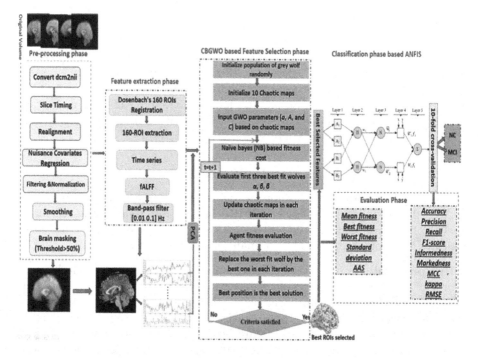

Fig. 1. General pipeline for the proposed model.

4   Experimental Results and Discussion

The proposed model was developed and tested via MATLAB (R2018b, The MathWorks, Inc.) on an Intel®CoreTM I5-6500 CPU 3.2 GHz processor and 16 GB RAM using Windows10 64-bit.

4.1   CBGWO-Based Feature Selection

Table 1 shows the best (BF) score, mean (μ), standard derivation (Std), number of selected features (SF), accuracy (ACC), and time consuming (CPU average time) obtained using BGWO based on CCM. We can see that, the optimal results (with the highest accuracy) are achieved using the following paramters setting; number of Agents (5), Max number of iterations (10), and Lower and Upper limits for binary search space [0,1].

The evidence of the convergence speed is provided in Fig. 2. The CBGWO algorithm has a strong robustness and has faster convergence speed to find the best solutions in less than 10 iterations. It is important to know how the kernel function decreases along the iterations and assist GWO to avoid local minima in the complex fMRI dataset.

Table 1. Results of the proposed BGWO model-based on NB.

Chaotic	BF	μ	Std	SF	AAS	Time/s	ACC
CCM	0.0371	0.0592	0.0153	82	0.513	31.000	0.8596

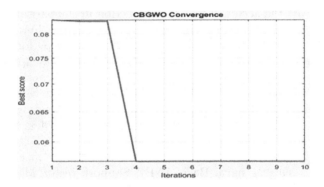

Fig. 2. BGWO algorithm convergence curve using CCB.

4.2 Diagnosis Based on ANFIS

Table 2 shows the accuracy results using the proposed model based on ANFIS and CCM with 10-fold cross validation (CV). We can see that, the highest accuracy achieved in almost all validation partitions with accuracy average ≈ 0.86. Moreover, Table 3 shos different statistical measures that used to validate the proposed model for MCI diagnosis. The accuracy achieved by Matthews correlation coefficient $MCC \approx 0.713$, $F1\text{-}score \approx 0.84$ which represents a perfect precision and recall even if the data is imbalanced, k is 0.69 which represent good agreement based on Kappa coefficient criteria $0.60 < k > 0.80$, $Info \approx 0.71$ which measures of how informed the model about positives and negatives, and $Mark \approx 0.73$ which measures trust worthiness of positive and negative predictions by the model.

Table 2. Accuracy results of the proposed CBGWO-ANFIS using 10-fold CV.

Ch.	CV_1	CV_2	CV_3	CV_4	CV_5	CV_6	CV_7	CV_8	CV_9	CV_{10}	Aver
CCM	0.8462	0.9167	1	0.7692	0.9231	1	0.6154	0.8333	1	0.6923	0.8596

Table 3. Results of CBGWO-ANFIS using different statistical measures.

Map	Acc	Recall	Precis.	F1-score	MCC	Kappa	Inform.	Mark.
CCM	0.8596	0.9257	0.7856	0.8395	0.7129	0.6900	0.7065	0.7273

In comparison with recent meta-heuristics optimization algorithms based on the performance of the selected features from rs-fMRI based MCI dataset. These algorithms are Whale optimization Algorithm (WOA) [13], Ant Lion Optimization (ALO) [15], Crow search optimization algorithm (CSA) [16], BGWO [17] algorithm, and BAT [18] algorithm. From Fig. 3, it is clearly seen that the CBGWO improved the accuracy for MCI prediction in comparison with WOA, ALO, BAT, CSA, and GWO algorithms with its chaotic maps. Further, since the BGWO-ANFIS is slightly better than CSA+Circle, we also performed 10 independent runs and afterward we used paired-sample t-test to compare these two algorithms. Results showed that CBGWO-ANFIS is significantly higher than CSA+Circle in the accuracy $(p = 0.03)$. From the results, we can see that, the proposed model achieved high prediction and quicker in locating the optimal solution. In general, it can search for the optimal solution in less than ten iterations.

Moreover, the proposed model compared with different machine learning (ML) techniques such as naïve Bayes (NB), Support vector Machine (SVM) with (Quadratic and Cubic), ANFIS, Logistic Regression (LR), Linear Discriminate Analysis (LDA), K-Nearest Neighbor (KNN), and Ensemble bagged Trees (EBT). All techniques are performed on the same MCI rs-fMRI dataset using 10-fold cross validation for training and testing sets. The results of these comparisons are shown in Fig. 4. It can be seen that, the more accurate results are achieved by using the proposed algorithm with high prediction accuracy ACC ($\approx$0.86), Sensitivity ($\approx$0.93), Specificity ($\approx$0.78), Precision ($\approx$0.79), and False positive rate (FPR) ($\approx$0.22), followed by the results achieved by ANFIS with ACC ($\approx$0.68), and followed by the results achieved by weighted KNN with ACC ($\approx$0.66). Further details about other methods can be seen in Fig. 4. In conclusion, the proposed model based ANFIS superior to conventional ML algorithms.

To validate the robustness of the proposed CBGWO-ANFIS model on other rs-fMRI dataset, we further examined the performance of the proposed model on a another dataset from ADNI [14]. Figure 5 shows the comparisons between CBGWO-ANFIS and several ML methods when being used ADNI dataset for classification of MCI patients from NC. As we can see, the proposed approach

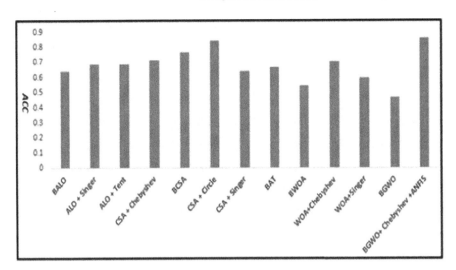

Fig. 3. Graphical representation accuracy results for different meta-heuristic algorithms.

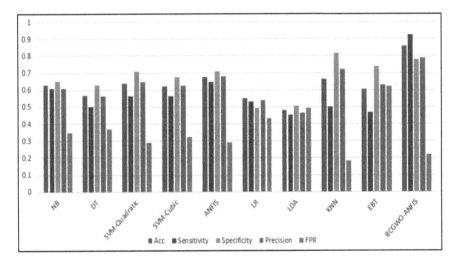

Fig. 4. Accuracy results for different ML techniques using different measures.

achieves the highest accuracy of 0.8232 and the highest sensitivity of 1. The specificity, precision and FPR of the proposed approach are also close to the best ones. These results further validate the better performance of the proposed CBGWO-ANFIS approach.

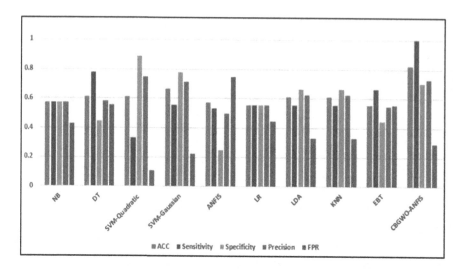

Fig. 5. Performance results and comparisons between CBGWO-ANFIS and other ML methods on ADNI dataset.

5   Conclusion and Future Work

In this paper, we proposed a novel MCI diagnosis model based on hybrid binary grey wolf optimizer, chaotic map and ANFIS algorithm (CBGWO-ANFIS). The experimental results demonstrate that the proposed hybrid model is reliable for MCI prediction especially with regard to high-dimensional data pattern recognition. The performance of the proposed model was measured on 127 subjects from real dataset. The experimental results show that the highest results obtained by the proposed model using 30 epochs and 10-fold cross validation. In a direct comparison, the proposed model exhibited better performance than bio-inspired optimization algorithms in literature. In future, the proposed model will be tested on more types of data (such as multimodal MRI), data from other centers, and even data of other brain diseases.

References

1. Vieira, S., Pinaya, W.H., Mechelli, A.: Using deep learning to investigate the neuroimaging correlates of psychiatric and neurological disorders: methods and applications. Neurosci. Biobehav. Rev. **74**, 58–75 (2017)
2. Suk, H.I., Wee, C.Y., Lee, S.W., Shen, D.: State-space model with deep learning for functional dynamics estimation in resting-state fMRI. NeuroImage **129**, 292–307 (2016)
3. Hosseini-Asl, E., Keynton, R., El-Baz, A.: Alzheimer's disease diagnostics by adaptation of 3D convolutional network. In 2016 IEEE International Conference on Image Processing (ICIP), pp. 126–130, September 2016

4. Han, X., Zhong, Y., He, L., Yu, P.S., Zhang, L.: The unsupervised hierarchical convolutional sparse auto-encoder for neuroimaging data classification. In: Guo, Y., Friston, K., Aldo, F., Hill, S., Peng, H. (eds.) BIH 2015. LNCS (LNAI), vol. 9250, pp. 156–166. Springer, Cham (2015). https://doi.org/10.1007/978-3-319-23344-4_16

5. Kim, J., Calhoun, V.D., Shim, E., Lee, J.H.: Deep neural network with weight sparsity control and pre-training extracts hierarchical features and enhances classification performance: Evidence from whole-brain resting-state functional connectivity patterns of schizophrenia. Neuroimage **124**, 127–146 (2016)

6. Grothe, M., Heinsen, H., Teipel, S.: Longitudinal measures of cholinergic forebrain atrophy in the transition from healthy aging to Alzheimer's disease. Neurobiol. Aging **34**(4), 1210–1220 (2013)

7. Li, Y., et al.: Abnormal resting-state functional connectivity strength in mild cognitive impairment and its conversion to Alzheimer's disease. Neural Plast. **2016**, 4680972 (2016). https://doi.org/10.1155/2016/4680972

8. Faris, H., Aljarah, I., Al-Betar, M.A., Mirjalili, S.: Grey wolf optimizer: a review of recent variants and applications. Neural Comput. Appl. **30**(2), 413–435 (2018)

9. Mirjalili, S., Mirjalili, S.M., Lewis, A.: Grey wolf optimizer. Adv. Eng. Softw. **69**, 46–61 (2014)

10. Li, T., Li, J., Liu, Z., Li, P., Jia, C.: Differentially private naive bayes learning over multiple data sources. Inf. Sci. **444**, 89–104 (2018)

11. Hasanipanah, M., Amnieh, H.B., Arab, H., Zamzam, M.S.: Feasibility of PSO-ANFIS model to estimate rock fragmentation produced by mine blasting. Neural Comput. Appl. **30**(4), 1015–1024 (2018)

12. Alam, S., Dobbie, G., Koh, Y.S., Riddle, P., Rehman, S.U.: Research on particle swarm optimization based clustering: a systematic review of literature and techniques. Swarm Evol. Comput. **17**, 1–13 (2014)

13. Anter, A., Gupta, D., Castillo, O.: novel parameter estimation in dynamic model via fuzzy swarm intelligence and chaos theory for faults in wastewater treatment plant. Soft Comput. 1–19 (2019). https://doi.org/10.1007/s00500-019-04225-7

14. ADNI. http://adni.loni.usc.edu/. Accessed April 2019

15. Mafarja, M., Eleyan, D., Abdullah, S., Mirjalili, S.: S-shaped vs. V-shaped transfer functions for ant lion optimization algorithm in feature selection problem. In: Proceedings of the International Conference on Future Networks and Distributed Systems, p. 14. ACM, July 2017

16. Anter, A., Ali, M.: Feature selection strategy based on hybrid crow search optimization algorithm integrated with chaos theory and fuzzy c-means algorithm for medical diagnosis problems. Soft Comput. 1–20 (2019)

17. Anter, A.M., Azar, A.T., Fouad, K.M.: Intelligent hybrid approach for feature selection. In: Hassanien, A.E., Azar, A.T., Gaber, T., Bhatnagar, R., F. Tolba, M. (eds.) AMLTA 2019. AISC, vol. 921, pp. 71–79. Springer, Cham (2020). https://doi.org/10.1007/978-3-030-14118-9_8

18. Rodrigues, D., et al.: A wrapper approach for feature selection based on bat algorithm and optimum-path forest. Expert Syst. Appl. **41**(5), 2250–2258 (2014)

Deep Learning via Fused Bidirectional Attention Stacked Long Short-Term Memory for Obsessive-Compulsive Disorder Diagnosis and Risk Screening

Chiyu Feng[1], Lili Jin[2], Chuangyong Xu[2], Peng Yang[1], Tianfu Wang[1], Baiying Lei[1(✉)], and Ziwen Peng[2,3(✉)]

[1] National-Regional Key Technology Engineering Laboratory for Medical Ultrasound, Guangdong Key Laboratory for Biomedical Measurements and Ultrasound Imaging, School of Biomedical Engineering, Health Science Center, Shenzhen University, Shenzhen 518060, China
leiby@szu.edu.cn

[2] College of Psychology and Sociology, Shenzhen University, Shenzhen 518060, China
pengzw@email.szu.edu.cn

[3] Department of Child Psychiatry, Shenzhen Kangning Hospital, Shenzhen University School of Medicine, Shenzhen, China

Abstract. The compulsive urges to perform stereotyped behaviors are typical symptoms of obsessive-compulsive disorder (OCD). OCD has certain hereditary tendencies and the direct OCD relatives (i.e., sibling (Sib)) have 50% of the same genes as patients. Sib has a higher probability of suffering from the same disease. Resting-state functional magnetic resonance imaging (R-fMRI) has made great progress by diagnosing OCD and identifying its high-risk population. Accordingly, we design a new deep learning framework for OCD diagnosis via R-fMRI data. Specifically, the fused bidirectional attention stacking long short-term memory (FBAS-LSTM) is exploited. First, we obtain two independent time series from the original R-fMRI by frame separation, which can reduce the length of R-fMRI sequence and alleviate the training difficulty. Second, we apply two independent BAS-LSTM learning on the hidden spatial information to obtain preliminary classification results. Lastly, the final diagnosis results are obtained by voting from the two diagnostic results. We validate our method on our in-house dataset including 62 OCD, 53 siblings (Sib) and 65 healthy controls (HC). Our method achieves average accuracies of 71.66% for differentiating OCD vs. Sib vs. HC, and outperforms the related algorithms.

Keywords: Obsessive-compulsive disorder diagnosis · Risk screening · Attention · LSTM · Fusion

This work was supported partly by National Natural Science Foundation of China (Nos. 31871113, 61871274, 61801305 and 81571758), National Natural Science Foundation of Guangdong Province (No. 2017A030313377), Guangdong Pearl River Talents Plan (2016ZT06S220), Shenzhen Peacock Plan (Nos. KQTD2016053112051497 and KQTD2015033016 104926), and Shenzhen Key Basic Research Project (Nos. JCYJ2017 0413152804728, JCYJ20180507184647636, JCYJ20170818142 347251 and JCYJ20170818094109846).

© Springer Nature Switzerland AG 2019
I. Rekik et al. (Eds.): PRIME 2019, LNCS 11843, pp. 34–43, 2019.
https://doi.org/10.1007/978-3-030-32281-6_4

1 Introduction

Obsessive-compulsive disorder (OCD) is a serious mental illness, which has incidence about 2% to 3%. Its typical symptoms are the intrusive thoughts and compulsive urges to perform stereotyped behaviors [1]. Up to now, the cause of OCD is still unknown, though some studies have shown that OCD is related to genetic factors [2] and the surrounding social environment [3]. Siblings (Sib) are the direct OCD relatives, which have 50% of the same OCD genes and live in a similar environment. Accordingly, Sib has the high risk of OCD. Recently, resting-state functional magnetic resonance imaging (R-fMRI) provides a non-invasive way to measure brain activity, which can assist physician to diagnose OCD and perform risk assessment of disease more effectively [4]. However, the diagnosis and risk assessment is not trivial even for health care professionals and it suffers from subjective human visual inspection as well.

To tackle these issues, there are many automatic diagnostic methods for clinical diagnosis of OCD. For example, Zhou *et al.* proposed a support vector machine (SVM) classification method based on whole brain volumetry and diffusion tensor imaging. Sen *et al.* proposed a method using Pearson's correlation scores for brain network construction to diagnose OCD. Lenhard *et al.* used four different machine learning methods to complete the early OCD diagnosis [5]. However, there are still many problems in the existing methods. First, the existing diagnostic framework does not take Sib into account as a high-risk population. Second, the traditional machine learning depends heavily on prior knowledge to select feature. Third, the traditional functional network construction needs complex pretreatment and computing.

To solve these problems, the recurrent neural network (RNN) and its variants have been widely used in the brain function analysis and the brain disease diagnosis via R-fMRI data [6]. For example, Yan *et al.* devised a new fully-connected bidirectional long short-term memory (FBi-LSTM) method to effectively learn the periodic brain status changes [7]. Dvornek *et al.* proposed a framework to diagnose depression using LSTM [8]. However, there is still no deep learning diagnostic model for OCD. Moreover, LSTM has limited processing power for long sequences such as R-fMRI.

Up to now, there are many studies devoted to solving this problem [9]. The attention model [9] is one of the most appealing methods, which effectively improves the limitations of LSTM for long sequence processing. In view of this, we devise a new deep learning framework for both OCD diagnosis and risk screening. Inspired by FBi-LSTM [7] and make further improvements, we first stack LSTM to extract deeper features. Then we enhance LSTM's processing of long sequences by adding attention module. Finally, we design a voting method called a fused bidirectional attention stacking LSTM (FBAS-LSTM) learning to solve the issue of LSTM's inadequate processing ability for long sequences. Experimental results on our own collected data from local hospital demonstrate the effectiveness of our proposed method, which achieve quite promising performance for both OCD diagnosis and risk screening.

2   Methodology

Figure 1 shows our framework based on FBAS-LSTM. We first divide bold signals into odd and even sequence via sampling interval. Then, we use two bidirectional attention stacked LSTM (BAS-LSTM) modules to extract the hidden information from the odd and even sequence, respectively. We simply fuse two modules in the output layer to get the final FBAS-LSTM module. In this work, we can input the forward and backward sequences into two separate BAS-LSTM modules. The final diagnosis results are obtained by 2 full connection (FC) layer and SoftMax (SF) function.

2.1   Data Preprocessing

The data was preprocessed using the Statistical Parametric Mapping toolbox (SPM8, www.fil.ion.ucl.ac.uk/spm), and Data Processing Assistant for Resting-State fMRI (DPARSFA version 2.2, http://rfmri.org/DPARSF). Image preprocessing consists of: (1) slice timing correction; (2) head motion correction; (3) realignment with the corresponding T1-volume; (4) nuisance covariate regression (six head motion parameters, white matte signal and cerebrospinal fluid signal); (5) spatial normalization into the stereotactic space of the Montreal Neurological Institute and resampling at 3×3 3 mm^3; (6) spatial smoothing with a 6-mm full-width half-maximum isotropic Gaussian kernel, and band-pass filtered (0.01–0.08 Hz). One OCD patient and two controls were excluded from further analysis because of a maximal absolute head motion larger than 1.5 mm or 1.5°.

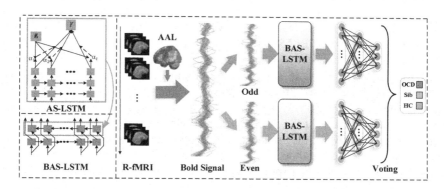

Fig. 1. The proposed FBAS-LSTM structure for OCD diagnosis.

2.2   FBAS-LSTM

LSTM is used to process time series [10]. LSTM cell includes the traditional RNN cell, each RNN cell receives last cell output (defined as h_{t-1}) and this input at the current time point (x_t). Then, each LSTM cell has a memory cell (defined as c_t) to save the history information and each cell adds a new input, which is the memory cell of last LSTM cell. In addition, there are three gates separately controlling input, forget and output. We define the original bold signal as $X = \{x_1, x_2, x_3 \ldots x_T\}$, where T is the number of time step, x_t is the signals strength. The LSTM cell is expressed as:

$$\text{Input gate}: i_t = \sigma(W_{xi}x_t + W_{hi}h_{t-1} + b_i), \tag{1}$$

$$\text{Forget gate}: f_t = \sigma(W_{xf}x_t + W_{hf}h_{t-1} + b_f), \tag{2}$$

$$\text{Output gate}: o_t = \sigma(W_{xo}x_t + W_{ho}h_{t-1} + b_o), \tag{3}$$

$$\text{Input modulation}: g_t = \varphi(W_{xc}x_t + W_{hc}h_{t-1} + b_c), \tag{4}$$

$$\text{Memory cell update}: c_t = i_t \odot g_t + f_t \odot c_{t-1}, \tag{5}$$

$$\text{Output}: h_t = o_t \odot \varphi(c_t) \tag{6}$$

Equation (1) denotes the input gate, which controls influence about current memory cell from x_t and h_{t-1}. Equation (2) defines the forget gate, which controls influence about current memory from previous memory cell c_{t-1}. Equation (3) is the forget gate, which controls influence about current memory from previous memory cell c_{t-1}. Equations (4) and (5) shows the process of memory cell update: getting the input modulation (defined as g_t) from x_t and h_{t-1} in Eq. (4) and updating memory cell with i_t, f_t and c_{t-1}. Finally, Eq. (6) shows the final output, which consists of current memory cell (c_t) and output gate (o_t). W, b, and σ and φ, and $\odot$ denote weight matrix, bias vector, sigmoid function, tanh function, and element-wise multiplication, respectively. Their subscripts represent the corresponding positions. We define the output of LSTM as $H = \{h_1, h_2, h_3 \ldots h_T\}$, where h_t is the feature of the region of interest (ROI) at t.

To further improve performance, we stack two layers of LSTM to extract deeper semantic information, and we referred to it as stack LSTM (S-LSTM). In fact, the output of S-LSTM can be used for further feature processing after fusion. However, S-LSTM can not reduce the dimension of the time-axis in the output of S-LSTM. Since not all the time step is equal to the representation of the disease status. Therefore, we introduce the attention model to weight feature on time-axis. The attention model is defined as:

$$u_t = \varphi(W_A h_t + b_A), \tag{7}$$

$$\alpha_t = \frac{exp(u_t^\top W_A)}{\sum_T exp(u_t^\top W_A)}, \tag{8}$$

$$Y = \sum_T \alpha_t h_t, \tag{9}$$

where $W_A, b_A, u_t, u_s, \alpha_t$ and Y denote the weight matrix, bias vector, a hidden representation of h_t, sequence level context vector, normalized importance weight and final output feature from the attention model. We refer to this S-LSTM as attention S-LSTM (AS-LSTM).

Since this model is bi-direction, both preceding and succeeding information from sequence can be effectively obtained via two opposite information from directions hidden layers. Then these two hidden layers are connected to one layer and we call this bidirectional AS-LSTM (BAS-LSTM). Here, we do not extract the label of SF output directly. We extract the scores of odd sequence and even sequence, respectively, and then average them. Specifically, we divide bold signal sequence into odd sequence and even sequence, which effectively shorten the sequence length without affecting the overall information. We can define the odd sequence and the even sequence as $X_o = \{x_1, x_3, x_5 \ldots x_{T-1}\}$ and $X_e = \{x_2, x_4, x_6 \ldots x_T\}$, and we fuse the two sequence by voting. Overall, we call this learning as FBAS-LSTM.

3 Experimental Setting and Results

3.1 Dataset and Implementation

In this paper, the dataset includes 62 OCD patients, 53 siblings (i.e., a high-risk OCD) and 65 HC. We use a 3.0-Tesla MR system (Philips Medical Systems, USA), which is equipped with an eight-channel phased-array head coil for data acquisition. The participants are instructed to relax with eyes closed, and stay awake without moving. Functional data are collected using gradient Echo-Planar Imaging (EPI) sequences (time repetition (TR) = 2000 ms; echo time (TE) = 60 ms; flip angle = 90°, 33 slices, field of view [FOV] = 240 mm × 240 mm, matrix = 64 × 64; slice thickness = 4.0 mm). For spatial normalization and localization, a high-resolution T1-weighted anatomical image was also acquired using a magnetization prepared gradient echo sequence (TR = 8 ms, TE = 1.7 ms, flip angle = 20°, FOV = 240 mm × 240 mm, matrix = 256 × 256, slice thickness = 1.0 mm).

In this work, the first layer of LSTM cell number is 65, the second is 35. The first layer of FC cell number is 50 and the second one is 30. Besides, we use root mean square prop and RMSprop as the optimizer [11] for speeding up training and categorical cross entropy as the loss function. The default setting in Keras are applied for learning rate, fuzz factor, rho, and learning rate decay. We set the batch size as 25. We use different performance metrics to evaluate the classification performance. For performance evaluation, we adopt the popular metrics such as accuracy (Acc), sensitivity (Sen), specificity (Spec), F1 score (F1), and balanced accuracy (BAC). Here, 5-fold evaluation is performed 5 times to avoid any bias which have 80% data as training set and 20% as testing set in each fold. All the experiments are conducted on a computer with GPU NVIDIA TITAN Xt and implemented using Keras library in Python.

3.2 Results

Firstly, the bidirectional LSTM (B-LSTM) result is used as baseline. The validity of the corresponding module is verified by comparing the experimental results of bidirectional attention LSTM (BA-LSTM), bidirectional stacked LSTM (BS-LSTM) and BAS-LSTM. For each model, we test four models separately: original sequence, odd sequence, even sequence and voting from odd and even sequence (Fused). The results are shown in Table 1 and the receivers operating curves (ROC) are shown in Fig. 2.

Table 1. Comparison of classification performance on different methods (%).

Method	Sequence	Acc	Sen	Spec	F1	BAC	AUC
B-LSTM	Original	47.22	47.25	73.67	46.22	60.46	64.32
	Odd	50.56	49.49	75.80	48.60	62.65	66.51
	Even	52.78	52.27	76.58	51.93	64.93	68.75
	Fused	55.00	54.77	77.93	53.70	66.35	70.84
BS-LSTM	Original	53.89	53.21	77.53	52.19	65.37	73.58
	Odd	60.56	59.54	80.74	59.44	70.14	76.99
	Even	63.33	62.27	81.89	62.08	72.07	80.37
	Fused	66.11	65.21	83.26	65.19	74.23	80.87
BA-LSTM	Original	57.78	56.52	79.27	56.42	67.89	76.39
	Odd	65.00	63.75	82.89	63.42	73.32	78.07
	Even	61.11	60.39	80.86	60.18	70.63	79.07
	Fused	67.22	66.23	84.14	65.65	75.19	82.42
BAS-LSTM	Original	62.78	61.54	81.67	61.49	71.61	81.12
	Odd	69.44	68.75	84.82	68.75	76.78	82.04
	Even	66.67	66.36	83.65	65.92	75.01	82.64
	Fused	**71.66**	**71.32**	**85.96**	**71.15**	**78.64**	**83.96**

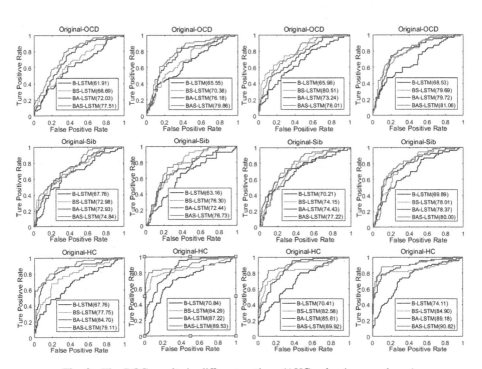

Fig. 2. The ROC results in different settings (AUC value in parentheses).

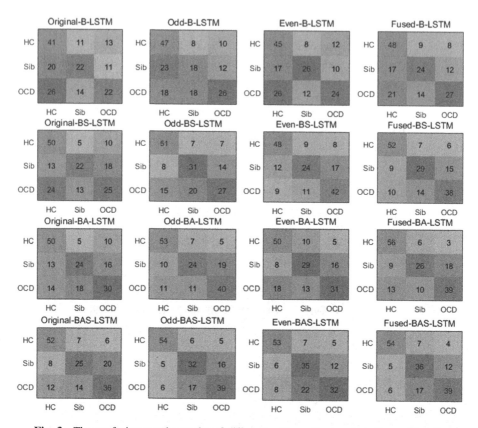

Fig. 3. The confusion matrix results of different settings (the y-axis is the real label).

We can observe that voting, stack and attention techniques can improve diagnosis performance. The main reasons are as follows. The method of separating point acquisition can effectively improve the diagnostic accuracy. Furthermore, compared with the analysis of single odd sequence or even sequence by BAS-LSTM, the analysis of both odd sequence and even sequence by FBAS-LSTM can avoid missing information when shortening the sequence and further improve the diagnostic performance. Besides, stack can increase the depth of the network and extract more advanced features, while attention can reduce the influence of long-time axis on LSTM training via weighting technique.

Then, the confusion matrices are shown in Fig. 3. As compared with HC group, Sib and OCD are found to have more classification errors. This suggests that brain activity of Sib and OCD is closer, which confirms our previous assumptions that Sib has the high risk of OCD.

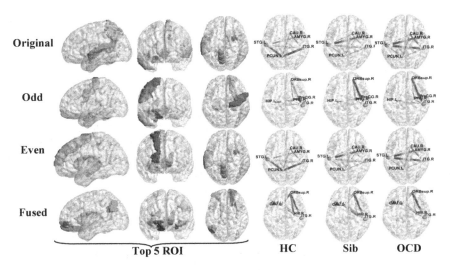

Fig. 4. Top 5 ROI that have greatest impact on the diagnosis of OCD, Sib and HC, and Top 5 ROI's functional connections of OCD, Sib and HC.

Finally, we identify the most relevant brain regions by counting the input gate weights of LSTM in the first layer. Specifically, we calculate the weights of input gates from all time points of LSTM at the first level, and the ROI corresponding to the five largest values of each weight vector. The top 5 important ROIs are obtained according to the number of occurrences of weights. For BAS-LSTM, we test four models separately: original sequence, odd sequence, even sequence and voting from odd and even sequence (Fused) and finally get four sets of top 5 ROIs. These ROIs are located in the brain as shown in Fig. 4. To better demonstrate the connections between these brain regions, we calculate the functional connections of these ROIs by using Pearson coefficient and present it in Fig. 4. We can see that OCD and SiB have more similar functional connections than HC. This further confirms our previous assumption that SIB has a high risk of OCD. Among them, Frontal_Sup_Orb_L, Frontal_Sup_Orb_R, Caudate_R and Caudate_L are the same as the results of existing brain functional network analysis between OCD and HC [12]. Our findings are consistent with the previous findings in [12].

Finally, we compare the performance of the proposed method with other existed methods. In order to make a fair comparison, we try to optimize the performance of other methods. Table 2 shows that our method can obtain higher accuracies than the

Table 2. Algorithm comparisons with other existing methods (%).

Method	Acc	Sen	Spec	F1	BAC	AUC
SVM	41.67	41.63	70.68	41.58	56.15	50.34
XGBOOST	51.67	51.34	75.79	51.32	63.57	68.07
CNN	59.44	59.98	79.93	58.11	69.96	76.33
FBAS-LSTM	**71.66**	**71.32**	**85.96**	**71.15**	**78.64**	**83.96**

existing methods. The mainly reason is that LSTM is more suitable for processing sequences and our method overcomes the weakness of traditional LSTM in long sequence processing

4 Conclusion

In this paper, we propose a new deep learning framework for OCD diagnosis based on FBAS-LSTM. We utilize the bold signal directly without the brain functional network construction, which can avoid the complex operation of constructing brain functional network. To the best of our knowledge, this is the first attempt for effective screening of high-risk groups while identifying OCD. We show its superiority and feasibility in a challenging OCD diagnosis and Sib screening. In addition, we find the brain regions related to OCD through the weight statistics of the model. Our findings are consistent with the existing clinical research results. In summary, the proposed model not only can effectively extract the information from ROI signal, but also improve OCD diagnostic accuracy and risk population screening.

References

1. Paula, B., et al.: Imbalance in habitual versus goal directed neural systems during symptom provocation in obsessive-compulsive disorder. Brain **138**, 798–811 (2015)
2. International, O.C.D.F.G.C., et al.: Revealing the complex genetic architecture of obsessive–compulsive disorder using meta-analysis. Mol. Psychiatry **23**, 1181–1188 (2017)
3. Howard, J., Serrano, W.C.: Anxiety, depression, and OCD: understanding common psychiatric conditions in the dermatological patient. In: França, K., Jafferany, M. (eds.) Stress and Skin Disorders, pp. 19–37. Springer, Cham (2017). https://doi.org/10.1007/978-3-319-46352-0_3
4. Tadayonnejad, R., et al.: Pregenual anterior cingulate dysfunction associated with depression in OCD: an integrated multimodal fMRI/1 H MRS study. Neuropsychopharmacology **43**, 1146–1155 (2018)
5. Lenhard, F., et al.: Prediction of outcome in internet-delivered cognitive behaviour therapy for paediatric obsessive-compulsive disorder: a machine learning approach. Int. J. Methods Psychiatr. Res. **27**, e1576 (2017)
6. Wang, H., et al.: Recognizing brain states using deep sparse recurrent neural network. IEEE Trans. Med. Imaging **38**(4), 1058–1068 (2018)
7. Yan, W., et al.: Deep chronnectome learning via full bidirectional long short-term memory networks for MCI diagnosis. Med. Image Comput. Comput. Assist. Intervention **2018**, 249–257 (2018)
8. Dvornek, N.C., Ventola, P., Pelphrey, K.A., Duncan, J.S.: Identifying autism from resting-state fMRI using long short-term memory networks. In: Wang, Q., Shi, Y., Suk, H.-I., Suzuki, K. (eds.) MLMI 2017. LNCS, vol. 10541, pp. 362–370. Springer, Cham (2017). https://doi.org/10.1007/978-3-319-67389-9_42
9. Yang, Z., et al.: Hierarchical attention networks for document classification. In: Proceedings of the 2016 Conference of the North American Chapter of the Association for Computational Linguistics: Human Language Technologies, pp. 1480–1489 (2016)

10. Hochreiter, S., Schmidhuber, J.: Long short-term memory. Neural Comput. **9**, 1735–1780 (1997)
11. Graves, A.: Generating sequences with recurrent neural networks. Computer Science (2013)
12. Xing, X., et al.: Diagnosis of OCD using functional connectome and Riemann kernel PCA. In: Medical Imaging 2019: Computer-Aided Diagnosis, vol. 10950, p. 109502C. International Society for Optics and Photonics (2019)

Modeling Disease Progression in Retinal OCTs with Longitudinal Self-supervised Learning

Antoine Rivail[1]([⊠]), Ursula Schmidt-Erfurth[1], Wolf-Dieter Vogel[1],
Sebastian M. Waldstein[1], Sophie Riedl[1], Christoph Grechenig[1], Zhichao Wu[2],
and Hrvoje Bogunovic[1]

[1] Christian Doppler Laboratory for Ophthalmic Image Analysis,
Department of Ophthalmology and Optometry, Medical University of Vienna,
Vienna, Austria
antoine.rivail@meduniwien.ac.at
[2] University of Melbourne, Melbourne, Australia

Abstract. Longitudinal imaging is capable of capturing the static anatomical structures and the dynamic changes of the morphology resulting from aging or disease progression. Self-supervised learning allows to learn new representation from available large unlabelled data without any expert knowledge. We propose a deep learning self-supervised approach to model disease progression from longitudinal retinal optical coherence tomography (OCT). Our self-supervised model takes benefit from a generic time-related task, by learning to estimate the time interval between pairs of scans acquired from the same patient. This task is (i) easy to implement, (ii) allows to use irregularly sampled data, (iii) is tolerant to poor registration, and (iv) does not rely on additional annotations. This novel method learns a representation that focuses on progression specific information only, which can be transferred to other types of longitudinal problems. We transfer the learnt representation to a clinically highly relevant task of predicting the onset of an advanced stage of age-related macular degeneration within a given time interval based on a single OCT scan. The boost in prediction accuracy, in comparison to a network learned from scratch or transferred from traditional tasks, demonstrates that our pretrained self-supervised representation learns a clinically meaningful information.

1 Introduction

Due to a rapid advancement of medical imaging, the amount of longitudinal imaging data is rapidly growing [1]. Longitudinal imaging is an especially effective observational approach, used to explore how disease processes develop over time in a number of patients, providing a good indication of disease progression. It enables personalized precision medicine [2] and it is a great source for automated image analysis. However, the automated modelling of disease progression

© Springer Nature Switzerland AG 2019
I. Rekik et al. (Eds.): PRIME 2019, LNCS 11843, pp. 44–52, 2019.
https://doi.org/10.1007/978-3-030-32281-6_5

faces many challenges: despite the large amount of data, associated human-level annotations are rarely available, which leads to several limitations in the current modelling methods. They are either limited to the annotated cross-sectional samples and miss most of the temporal information. Or, they are based on known handcrafted features and simplified models of low complexity.

To overcome these limitations, we propose a solution based on self-supervised learning. Self-supervised learning consists of learning an auxiliary or a so-called *pretext* task on a dataset without the need for human annotations to generate a generic representation. This representation can then be transferred to solve complex supervised tasks with a limited amount of data. We propose a pretext task that exploits the availability of large numbers of unlabelled longitudinal images, focuses on learning temporal-specific patterns and that is not limited by irregular time-sampling or lack of quality in image registration, which are common issues in longitudinal datasets. The learnt representation is compact and allows for transfer learning to more specific problems with limited amount of data. We demonstrate the capability of our proposed method on a longitudinal retinal optical coherence tomography (OCT) dataset of patients with early/intermediate age-related macular degeneration (AMD).

Clinical Background. In the current ophthalmic clinical practice, optical coherence tomography (OCT) is the most commonly used retinal imaging modality. It provides 3-dimensional in-vivo information of the (pathological) retina with a micrometer resolution. Typically, volumetric OCTs are rasterized scans, where each *B-Scan* is a cross-sectional image of the retinal morphology. AMD is a major epidemic among the elderly and advanced stage of AMD is the most common cause of blindness in industrialized countries, with two identified main forms: geographic atrophy (GA) and choroidal neovascularization (CNV). The progression from early or *intermediate*, symptomless stages of AMD to *advanced* stage is extremely variable between patients and very difficult to estimate clinically. Robust and accurate prediction at the individual patient level is a critically important medical need in order to manage or prevent irreversible vision loss.

Related Work. Current analysis of clinical data and disease course development evolves around traditional statistical approaches, where time-series models are fit to a limited amount of known biomarkers describing the disease status [2, 3]. Such models are problematic in case of a disease such as AMD where the underlying mechanisms are still poorly understood and main biomarkers are yet undiscovered [4]. Disease progression modelling from longitudinal imaging data has been most active in the field of Neuroimaging for modeling the progression of Alzheimer's disease, largely due to the public availability of a longitudinal brain magnetic resonance images under Alzheimer's Disease Neuroimaging Initiative (ADNI). There, a variety of regression-based methods have been applied to fit logistic or polynomial functions to the longitudinal dynamic of each imaging biomarker [5]. Other efforts have been focusing on non-parametric Gaussian process (GP) models [6] but the specification of the joint covariance structure of the image features to account for spatial and temporal correlation has been found to still be computationally prohibitive. In addition, these methods are often linear, and/or treat the data as cross-sectional, and thus do not exploit nonlinear relationships. Self-supervised learning was already successfully applied to time-series video data in the field of computer vision, where Lee et al. developed a solution based on time-shuffling [7], which inspired our method.

Contribution. We propose a novel self-supervised task, suited for learning a compact representation of longitudinal imaging data that captures time-specific patterns without the need of segmentation or a priori information. We assume that the information of the future evolution and disease progression is encoded in an observed series of images to a certain degree. Hence, we train our model on a pretext task: estimating the time interval between pairs of images of the same patient. Thus, an implicit aging model is built resulting in a compact representation that contains knowledge about healthy and disease evolution. As this task does not rely on annotations, perfect registration or regular sampling intervals, we are able to incorporate large unlabelled longitudinal datasets without the need of time and cost intensive pre-processing or annotation generation. We demonstrate that the model is able to learn the given pretext task and that it is capable of capturing the longitudinal evolution. Furthermore, we show that such a representation can be transferred to other longitudinal problems such as a prediction or survival estimation setting with limited amount of training data. In our case, we predict the future conversion to advanced stage of AMD within a certain time interval. In contrast to a model learnt from scratch or transferred from non longitudinal task (ex. autoencoder), we observe a boost in accuracy when fine-tuning a model trained on our new pretext task.

2   Self-supervised Learning of Spatio-Temporal Representations

In this Section, we will present the pretext task that we chose for self-supervised learning of longitudinal imaging data, the deep networks that we implemented to solve it, and finally how we extract the representations to transfer them to different problems.

Self-supervised Learning Paradigm. To learn the pretext task, we train a deep Siamese network (Fig. 1) [8]. The Siamese structure reflects the symmetry of the problem and allows to train a single encoder, which can be later transferred either as a fixed feature extractor or as a pretrained network for fine-tuning.

Let X_{t_1}, X_{t_2} be two OCT images acquired from the same patient, and t_1, t_2 be the corresponding time-points of acquisition. These images are encoded into a compact representation (H_{t_1}, H_{t_2}) by the **encoder network**, F. The **pair-interval network**, G, predicts the time interval or relative time difference between the pair of B-Scans, $\Delta T = t_2 - t_1$. Note that the order of the pair do not need to be chronological.

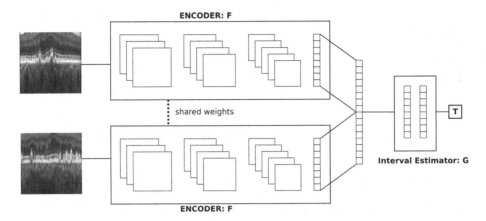

Fig. 1. Deep siamese network learning the self supervised task. It consists of predicting the acquisition time interval separating a pair of OCT scans of the same patient. Each B-Scan is encoded by the **encoder**, then the codes are used as input for the **pair-interval** network, which outputs the estimated time interval.

$$F: \boldsymbol{X}_t \to \boldsymbol{H}_t \tag{1}$$

$$G: [\boldsymbol{H}_{t_1}, \boldsymbol{H}_{t_2}] \to \widehat{\boldsymbol{\Delta T}} \tag{2}$$

$$\text{Loss} = ||\, G([\boldsymbol{H}_{t_1}, \boldsymbol{H}_{t_2}]) - \boldsymbol{\Delta T}||_2 \tag{3}$$

The entire siamese network is trained by minimizing the L_2 loss of this regression task. After the training, the encoder network can be transferred to extract features for other tasks.

Implementation. The encoder is implemented as a deep convolutional network, with three blocks of three layers (each layer: 3×3 convolution layer with batch normalization and ReLU activation with 16, 32, 64 channels for block 1, 2, 3) with a max pooling layer at the end of each block. The last block is followed by a fully connected layer (128 units), which outputs the encoded version of the B-scan (denoted as **vgg**). We also tested a version with skip connections followed by concatenation between the blocks (denoted as **dense**). The pair-interval network has two fully connected layers and outputs the estimated time interval.

Learning Setup. The network is trained by minimizing L2 loss with gradient-descent algorithm Adam [9]. We trained the network for 600k steps and computed validation loss every 12k steps. We kept the model with highest validation loss.

Transfer of Representations. The representations, $\mathbf{H}_t$, extracted from the encoder network, are used as input for a classification problem. This transfer allows to evaluate whether these representations are containing meaningful information regarding the patient-specific evolution of AMD. We directly transfer the trained encoder from the deep siamese network to a classification task by adding

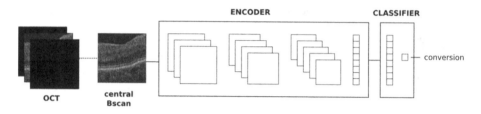

Fig. 2. Transfer learning of the trained encoder. A classification block is added at the end of the encoder and the network is fine-tuned on the new problem.

a final block to perform classification (Fig. 2). The classification block consists of two fully connected layers, the first one with ReLU activation, the last one with softmax. The resulting network is fine-tuned by minimizing cross-entropy.

3 Experiments and Results

Here, we provide details about the self-supervised training and its evaluation with respect to the time prediction. Then, we transfer the representation obtained from self-supervised training to a classification task, where we predict from longitudinal retinal OCT the conversion from intermediate to advanced AMD within different time intervals.

3.1 Dataset

The longitudinal dataset used for training and validation contains 3308 OCT scans from 221 patients (420 eyes) diagnosed with intermediate AMD. Follow-up scans were acquired in a three or six months interval up to 7 years, and were included in the dataset up to the time-point of conversion to advanced AMD. Follow-up acquisitions were automatically registered by scanner software (Spectralis OCT®, Heidelberg Engineering, GER). Within this population, 48 eyes converted to GA, an advanced stage of AMD.

Cross-Validation. The patients are divided in 6 fixed folds to perform cross-validation. For the pretext task and the prediction of conversion, we used one fold as test, one as validation and the remaining folds as training data. The pre-training with self-supervised learning use the same training sets as the prediction of conversion.

Preprocessing. First, the bottom-most layer of the retina, the Bruch's membrane (BM), was segmented using [10], followed by a flattening of the concave structure and an alignment of the BM over all scans. Finally, scans were cropped to the same physical field of view (6 mm × 0.5 mm) and resampled to 128 × 128.

3.2 Learning the Pretext Task

The success of learning the pretext task was evaluated using R^2 and the mean absolute error (MAE) of the *interval* prediction. In addition, we verified how

Table 1. Evaluation of interval prediction (R^2 and MAE), and of sample order prediction (accuracy) in the pretext task obtained from a 6-fold cross-validation.

Network	R^2	MAE (months)	Accuracy
vgg	**0.566**	**7.69**	**0.843**
dense	0.505	8.35	0.823

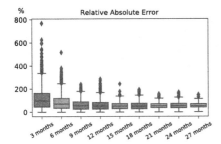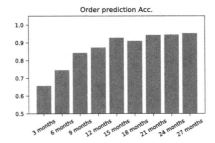

Fig. 3. *Left*, boxplot of relative absolute error [%] for the different time intervals in the dataset (3 to 24 months). *Right*, sample order prediction accuracy for the different time intervals in the dataset.

well the network could predict the temporal *order* of samples, which is done by evaluating the accuracy of predicting the correct sign of the interval. To obtain a volume prediction (the network is trained on B-Scans), we took the mean of all scan predictions. The best performance was achieved by the vgg-like model with a R^2 of **0.566**, a MAE of **7.69 months** and an accuracy for order prediction of **0.843** (Table 1). Figure 3 displays MAE and order prediction accuracy for the different time intervals (the visit intervals are roughly a multiple of three months). We observed that the network was able to predict the order, even for the smallest interval (3 months) with an accuracy of **0.66** (a random performance yielding 0.5). The absolute time-interval regression error was greater for large intervals (Fig. 3), with a tendency to underestimate the interval, probably because of a uniform distribution of training intervals centered on zero. However the relative error was decreasing for larger intervals. These results show that it is possible to estimate the time interval between two OCTs to a certain extent, which allows to learn a generic evolution model for the retina. In the next experiment, we verified that this model contained relevant longitudinal information to solve a specific clinical prediction task.

3.3 Conversion to Advanced AMD Classification

We applied the representation on a binary classification task, where we predicted from a single OCT representation, whether a patient eye will convert to GA within defined intervals of 6 months, 12 and 18 months (three separated binary problems). For the evaluation dataset we used the same 6 folds and their train,

Table 2. Conversion classification of patients suffering from intermediate AMD. We performed a 6-fold cross-validation and display the ROC AuC and average precision (mean and standard deviation) for three settings: conversion within 6 months (m.), 12 months and 18 months.

Model	ROC AuC		
	6 m.	12 m.	18 m.
Training from scratch (i)	0.640 ± 0.067	0.651 ± 0.076	0.676 ± 0.095
OCT autoencoder (ii)	0.650 ± 0.144	0.519 ± 0.060	0.677 ± 0.088
Self-supervised (ours)	**0.753**± 0.061	**0.784** ± 0.067	**0.773** ± 0.074
Model	Average precision		
	6 m.	12 m.	18 m.
Training from scratch (i)	0.309 ± 0.114	0.282 ± 0.0125	0.300 ± 0.107
OCT autoencoder (ii)	0.277 ± 0.152	0.283 ± 0.144	0.329 ± 0.117
Self-supervised (ours)	**0.367** ± 0.084	**0.394** ± 0.115	**0.463** ± 0.133

validation and test subsets. However, we included only one OCT per patient, in order to simulate a single visit. For patients who converted we chose the acquisition having the largest distance to conversion within the given interval. For patients showing no conversion within the study we chose the acquisition within the given interval with the furthest distance to the last patients acquisition. For each OCT volume, we used the trained encoder to extract a representation vector from the central B-Scan.

We restricted the input to a single time-point to verify that the encoded information allows to evaluate directly the stage of the patient. We fine-tuned the encoder on each fold and kept the epoch with the highest validation loss. We tested two different baselines with the same structure as the model transferred from the self-supervised learning, (i) trained from scratch (no transfer), and (ii) transferred from an autoencoder trained on cross-sectional OCTs with mean square error as reconstruction loss. The Autoencoder baseline allows to verify that our method is learning specific longitudinal features. The networks pretrained with our method or with an autoencoder are fine-tuned using ADAM optimizer by minimizing cross-entropy. After 20 epochs, the best epoch is selected using validation AuC. We performed a grid search on the learning rate, number of features in the classification block, and dropout rate, each setting is repeated 5 times. The best hyperparameter combination was chosen based on the best average validation AuC. The final cross-validated test performance was evaluated using ROC AuC and average precision. We observed that all the networks rapidly overfit on the dataset, which was expected given the size of the dataset (around 260 training samples) and the high capacity of the network. For all intervals, the self-supervised method shows best performance for both ROC AuC and average precision. On the other hand, transfer from the OCT autoencoder is only marginally better than the network trained from scratch (Table 2). The difference between the transfer from OCT autoencoder and our new self-supervised task shows that the latter captures longitudinal information, which is not available in the autoencoder. Our method allows to quickly train a deep network on this challenging task with a small number of annotations.

4   Discussion and Conclusions

Effective modeling of disease progression from longitudinal data has been a long pursued goal in medical image analysis. We presented a method based on self-supervised learning, which builds an implicit evolution model by taking benefit from longitudinal unlabelled data. This method allows to build representations in an unsupervised way that captures time-specific patterns in the data. The representation can be transferred to solve many longitudinal problems, such as patient-specific early prediction or risk estimation. Unlike reconstruction based methods, the pretext task allows to train on irregular longitudinal data, with irregular time-sampling or limited anatomical registration. The trained encoder can be transferred easily to related longitudinal problems with limited amount of annotated data. In this paper, we applied the method on longitudinal OCTs of patients with intermediate AMD. The learned features were successfully transferred to the problem of predicting incoming disease onset to advanced AMD. There are, however, some limitations in the proposed method. The method is trained on single B-scans instead on the full volume, which highly reduces the memory footprint, but introduces some intermediate steps to generate a patient representation and makes the pretext task harder, as the evolution of each OCT volume might not be uniformly distributed. Although we demonstrated the capability of our approach on retinal OCT scans, the method is not limited to this imaging modality or anatomical region, and may be applied to other longitudinal medical imaging datasets as well.

References

1. Fujimoto, J., Swanson, E.: The development, commercialization, and impact of optical coherence tomography. Invest. Ophthalmol. Vis. Sci. **57**(9), OCT1–OCT13 (2016)
2. Vogl, W.D., Waldstein, S.M., Gerendas, B.S., Schlegl, T., Langs, G., Schmidt-Erfurth, U.: Analyzing and predicting visual acuity outcomes of anti-VEGF therapy by a longitudinal mixed effects model of imaging and clinical data. Invest. Ophthalmol. Vis. Sci. **58**(10), 4173 (2017)
3. Vogl, W.D., Waldstein, S.M., Gerendas, B.S., Schmidt-Erfurth, U., Langs, G.: Predicting macular edema recurrence from spatio-temporal signatures in optical coherence tomography images. IEEE Trans. Med. Imaging **36**(9), 1773–1783 (2017)
4. Schmidt-Erfurth, U., et al.: Prediction of individual disease conversion in early AMD using artificial intelligence. Invest. Ophthalmol. Vis. Sci. **59**(8), 3199–3208 (2018)
5. Sabuncu, M.R., Bernal-Rusiel, J.L., Reuter, M., Greve, D.N., Fischl, B., Alzheimer's Disease Neuroimaging Initiative: Event time analysis of longitudinal neuroimage data. NeuroImage **97**, 9–18 (2014)

6. Lorenzi, M., Ziegler, G., Alexander, D.C., Ourselin, S.: Efficient Gaussian process-based modelling and prediction of image time series. In: Ourselin, S., Alexander, D.C., Westin, C.-F., Cardoso, M.J. (eds.) IPMI 2015. LNCS, vol. 9123, pp. 626–637. Springer, Cham (2015). https://doi.org/10.1007/978-3-319-19992-4_49

7. Lee, H., Huang, J., Singh, M., Yang, M.: Unsupervised representation learning by sorting sequences. CoRR abs/1708.01246 (2017)

8. Bromley, J., Guyon, I., LeCun, Y., Säckinger, E., Shah, R.: Signature verification using a "siamese" time delay neural network. In: Advances in NIPS, pp. 737–744 (1994)

9. Kingma, D.P., Ba, J.: Adam: A method for stochastic optimization. arXiv:1412.6980 (2014)

10. Chen, X., Niemeijer, M., Zhang, L., Lee, K., Abramoff, M.D., Sonka, M.: Three-dimensional segmentation of fluid-associated abnormalities in retinal OCT: probability constrained graph-search-graph-cut. IEEE Trans. Med. Imaging 31(8), 1521–1531 (2012)

Predicting Response to the Antidepressant Bupropion Using Pretreatment fMRI

Kevin P. Nguyen[(⊠)], Cherise Chin Fatt, Alex Treacher, Cooper Mellema, Madhukar H. Trivedi, and Albert Montillo

University of Texas Southwestern Medical Center, Dallas, USA
kevin3.nguyen@utsouthwestern.edu

Abstract. Major depressive disorder is a primary cause of disability in adults with a lifetime prevalence of 6–21% worldwide. While medical treatment may provide symptomatic relief, response to any given antidepressant is unpredictable and patient-specific. The standard of care requires a patient to sequentially test different antidepressants for 3 months each until an optimal treatment has been identified. For 30–40% of patients, no effective treatment is found after more than one year of this trial-and-error process, during which a patient may suffer loss of employment or marriage, undertreated symptoms, and suicidal ideation. This work develops a predictive model that may be used to expedite the treatment selection process by identifying for individual patients whether the patient will respond favorably to bupropion, a widely prescribed antidepressant, using only pretreatment imaging data. This is the first model to do so for individuals for bupropion. Specifically, a deep learning predictor is trained to estimate the 8-week change in Hamilton Rating Scale for Depression (HAMD) score from pretreatment task-based functional magnetic resonance imaging (fMRI) obtained in a randomized controlled antidepressant trial. An unbiased neural architecture search is conducted over 800 distinct model architecture and brain parcellation combinations, and patterns of model hyperparameters yielding the highest prediction accuracy are revealed. The winning model identifies bupropion-treated subjects who will experience remission with the number of subjects needed-to-treat (NNT) to lower morbidity of only 3.2 subjects. It attains a substantially high neuroimaging study effect size explaining 26% of the variance ($R^2 = 0.26$) and the model predicts post-treatment change in the 52-point HAMD score with an RMSE of 4.71. These results support the continued development of fMRI and deep learning-based predictors of response for additional depression treatments.

Keywords: Depression · Treatment response · fMRI · Neuroimaging · Deep learning · Neural architecture search

© Springer Nature Switzerland AG 2019
I. Rekik et al. (Eds.): PRIME 2019, LNCS 11843, pp. 53–62, 2019.
https://doi.org/10.1007/978-3-030-32281-6_6

1 Introduction

Major depressive disorder (MDD) has a lifetime prevalence of 6–21% worldwide and is a major cause of disability in adults [12]. Though half of MDD cases are treated with medication, there are dozens of antidepressants available and a patient's response to each is highly unpredictable [7]. The current standard in healthcare entails a long trial-and-error process in which a patient tries a series of different antidepressants. The patient must test each drug for up to 3 months, and if satisfactory symptomatic improvement is not achieved within this time, the clinician modifies the dosage or selects a different drug to test next. This trial-and-error process may take months to years to find the optimal treatment, during which patients suffer continued debilitation, including worsening symptoms, social impairment, loss of employment or marriage, and suicidal ideation. It has been shown that 30–40% of patients do not find adequate treatment after a year or more of drug trials [19,22]. Consequently, a predictive tool that helps prioritize the selection of antidepressants that are best suited to each patient would have high clinical impact.

This work demonstrates the use of deep learning and pretreatment task-based fMRI to predict long-term response to bupropion, a widely used antidepressant with a response rate of 44% [15]. An accurate screening tool that distinguishes bupropion responders from non-responders using pretreatment imaging would reduce morbidity and unnecessary treatment for non-responders and prioritize the early administration of bupropion for responders.

The use of functional magnetic imaging (fMRI) measurements to infer quantitative estimates of bupropion response is motivated by evidence for an association between fMRI and antidepressant response. For example, resting-state activity in the anterior cingulate cortex as well as activity evoked by reward processing tasks in the anterior cingulate cortex and amygdala have all been associated with antidepressant response [13,16,17].

In this work, predictive models of individual response to bupropion treatment are built using deep learning and pretreatment, task-based fMRI from a cohort of MDD subjects. The novel contributions of this work are: (1) the first tool for accurately predicting long-term bupropion response, and (2) the use of an unbiased neural architecture search (NAS) to identify the best-performing model and brain parcellation from 800 distinct model architecture and parcellation combinations.

2 Methods

2.1 Materials

Data for this analysis comes from the EMBARC clinical trial [23], which includes 37 subjects who were imaged with fMRI at baseline and then completed an 8-week trial of bupropion XL. To track symptomatic outcomes, the 52-point Hamilton Rating Scale for Depression (HAMD) was administered at baseline and week 8 of antidepressant treatment. Higher HAMD scores indicate greater

MDD severity. Quantitative treatment response for each subject was defined as ΔHAMD = HAMD(week 8) $-$ HAMD(baseline), where a negative ΔHAMD indicates improvement in symptoms. The mean ΔHAMD for these subjects was -5.98 ± 6.25, suggesting a large variability in individual treatment outcomes. For comparison, placebo-treated subjects in this study exhibited a mean ΔHAMD of -6.70 ± 6.93.

Image Acquisition. Subjects were imaged with resting-state and task-based fMRI (gradient echo-planar imaging at 3T, TR of 2000 ms, $64 \times 64 \times 39$ image dimensions, and $3.2 \times 3.2 \times 3.1$ mm voxel dimensions). Resting-state fMRI was acquired for 6 min. Task-based fMRI was acquired immediately afterwards for 8 min during a well-validated block-design reward processing task assessing reactivity to reward and punishment [8,11]. In this task, subjects must guess in the *response phase* whether an upcoming number will be higher or lower than 5. They are then informed in the *anticipation phase* if the trial is a "possible win", in which they receive a \$1 reward for a correct guess and no punishment for an incorrect guess, or a "possible loss", in which they receive a -\$0.50 punishment for an incorrect guess and no reward for a correct guess. In the *outcome phase*, they are then presented with the number and the outcome of the trial.

2.2   Image Preprocessing

Both resting-state and task-based fMRI images were preprocessed as follows. Frame-to-frame head motion was estimated and corrected with FSL MCFLIRT, and frames where the norm of the fitted head motion parameters was >1 mm or the intensity Z-score was >3 were marked as outliers. Images were then skull-stripped using a combination of FSL BET and AFNI Automask. To perform spatial normalization, fMRI images were registered directly to an MNI EPI template using ANTs. This coregistration approach has been shown to better correct for nonlinear distortions in EPI acquisitions compared to T1-based coregistration [2,6]. Finally, the images were smoothed with a 6 mm Gaussian filter.

Predictive features were extracted from the preprocessed task-based fMRI images in the form of contrast maps (i.e. spatial maps of task-related neuronal activity). Each task-based fMRI image was fit to a generalized linear model,

$$\boldsymbol{Y} = \boldsymbol{X} \times \boldsymbol{\beta} + \epsilon$$

where $\boldsymbol{Y}$ is the *time $\times$ voxels* matrix of BOLD signals, $\boldsymbol{X}$ is the *time $\times$ regressors* design matrix, $\boldsymbol{\beta}$ is the *regressors $\times$ voxels* parameter matrix, and ϵ is the residual error, using SPM12. The design matrix $\boldsymbol{X}$ was defined as described in [11] and included regressors for the response, anticipation, outcome, and inter-trial phases of the task paradigm. In addition, a reward expectancy regressor was included, which had values of $+0.5$ during the anticipation phase for "possible win" trials and -0.25 during the anticipation phase for "possible loss" trials. These numbers correspond to the expected value of the monetary reward/punishment in each trial. In addition to these task-related regressors and their first temporal derivatives, the head motion parameters and outlier frames were also included as regressors in $\boldsymbol{X}$.

After fitting the generalized linear model, contrast maps for anticipation ($\boldsymbol{C}_{antic}$) and reward expectation ($\boldsymbol{C}_{re}$) were computed from the fitted $\boldsymbol{\beta}$ coefficients:

$$\boldsymbol{C}_{antic} = \boldsymbol{\beta}_{\text{anticipation}} - \boldsymbol{\beta}_{\text{inter-trial}}$$
$$\boldsymbol{C}_{re} = \boldsymbol{\beta}_{\text{reward expectation}}$$

To extract region-based features from these contrast maps, three custom, study-specific brain parcellations (later referred to as *ss100*, *ss200* and *ss400*) were generated with 100, 200, and 400 regions-of-interest (ROIs) from the resting-state fMRI data using a spectral clustering method [5]. Each parcellation was then used to extract mean contrast values per ROI. The performance achieved with each of these custom parcellations, as well as a canonical functional atlas generated from healthy subjects (Schaefer 2018, 100 ROIs) [20], is compared in the following experiments.

2.3 Construction of Deep Learning Predictive Models

Dense feed-forward neural networks were constructed to take the concatenated ROI mean values from the two contrast maps as inputs and predict 8-week ΔHAMD. Rather than hand-tuning model hyperparameters, a random search was conducted to identify a high-performing model for predicting response to bupropion. The random search is an unbiased neural architecture search (NAS) that was chosen because it has been shown to outperform grid search [1] and when properly configured can provide performance competitive with leading NAS methods such as ENAS [14].

200 architectures were sampled randomly from a uniform distribution over a defined hyperparameter space (Table 1) and then used to construct models that were trained in parallel on 4 NVIDIA P100 GPUs. All models contained a single neuron output layer to predict ΔHAMD and were trained with the *Nadam* optimizer, 1000 maximum epochs, and early stopping after 50 epochs without decrease in validation root mean squared error (RMSE).

The combination of 200 model architectures with 4 different parcellations resulted in a total of 800 distinct model configurations that were tested. To ensure robust model selection and to accurately estimate generalization performance, these 800 model configurations were tested with a nested K-fold cross-validation scheme with 3 outer and 3 inner folds. Although a single random split is commonly used in place of the outer validation loop, a nested cross-validation ensures that no test data is used during training or model evaluation and provides an unbiased estimate of final model performance [24]. Within each outer fold, the best-performing model was selected based on mean root mean squared error (RMSE) over the inner folds. The model was then retrained on all training and validation data from the inner folds and final generalization performance

Table 1. Hyperparameter space defined for the random neural architecture search. For each model, one value was randomly selected from each of the first set of hyperparameters; for each layer in each model, one value was randomly selected from the second set of hyperparameters.

Hyperparameter	Possible values
Per-model hyperparameters	
Number of dense hidden layers	1, 2, 3, 4, 5
Number of neurons in 1st hidden layer	$32N$ for $N \in [1, \ldots, 16]$
Activation for all layers	Leaky ReLU, ReLU, ELU, PReLU
Learning rate	$0.0001n$ for $n \in [1, \ldots, 50]$
Per-layer hyperparameters	
% decrease in neurons from previous layer	None, 0.25, 0.5, 0.75
Weight regularization	L_1, L_2, L_1 and L_2
Activity regularization	L_1, L_2, L_1 and L_2
Batch normalization	Yes, No
Dropout rate	0, 0.3, 0.5, 0.7

was evaluated on the held-out test data of the outer fold. Repeating this process for each outer fold yielded 3 best-performing models, and the mean test performance of these models is reported here.

3   Results and Discussion

3.1   Neural Architecture Search (NAS)

Results indicate that the NAS is beneficial. In particular, a wide range of validation RMSE was observed across the 800 tested model configurations (Fig. 1). Certain models performed particularly well achieving RMSE approaching 4.0, while other model architectures were less suitable. NAS helped identify high-performing configurations expediently.

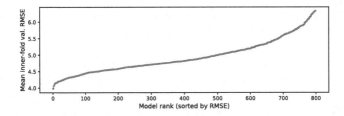

Fig. 1. Mean inner validation fold RMSE of the 800 model architecture & parcellation combinations evaluated in the unbiased neural architecture search. Results from one outer cross-validation fold are illustrated here, and findings for the other two folds were similar.

The information from the NAS can be examined for insight into what configurations constitute high versus low performing models and whether the ranges of hyperparameters searched were sufficiently broad. Towards this end, the hyperparameter distributions of the top and bottom quartiles of these 800 model configurations, sorted by RMSE, were compared. Substantial differences in the hyperparameter values that yielded high and low predictive accuracy are observed (Fig. 2). Notably, the custom, study-specific parcellation with 100 ROIs (*ss100*) provided significantly better RMSE than the "off-the-shelf" Schaefer parcellation ($p = 0.023$). Additionally, the top quartile of models using *ss100* used fewer layers (1–2), but more neurons (384–416) in the first hidden layer, compared to the bottom quartile of models. Note that unlike in a parameter sensitivity analysis, where ideal results exhibit a uniform model performance over a wide range of model *parameters*, in a neural architecture search, an objective is to demonstrate adequate coverage over a range of *hyperparameters*. This objective is met when local performance maxima are observed. This is shown in (Fig. 2b, c and d) where peaks in the top quartile (blue curve) of model architectures are evident.

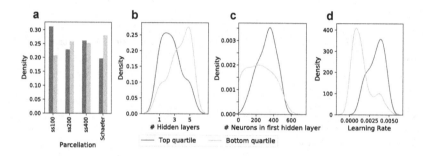

Fig. 2. Hyperparameter patterns for the top (blue) and bottom (orange) quartiles of the 800 model configurations evaluated in the unbiased neural architecture search. Representative results for one of the outer cross-validation folds are presented. a: Top quartile models tended to use the *ss100* parcellation, while bottom quartile models tended to use the Schaefer parcellation. b–d: Distributions of three selected hyperparameters compared for the top and bottom quartiles of model configurations, revealing the distinct patterns of hyperparameters for high-performing models. The top quartile of model architectures have fewer layers (peaking at 1–2) but more neurons in the first hidden layer (peaking at 384–416 neurons). (Color figure online)

The best performing model configuration used an architecture with two hidden layers and the 100-ROI study-specific parcellation (*ss100*). Regression accuracy in predicting ΔHAMD in response to bupropion treatment was RMSE 4.71 and R^2 0.26. This R^2 value (95% confidence interval 0.12–0.40 for $n = 37$) constitutes a highly significant effect size for a neuroimaging study where effect sizes are commonly much lower, e.g. 0.01–0.10 in [3] and 0.09–0.15 in [21]. Furthermore, this predictor identifies individuals who will experience clinical remission

Table 2. Performance of the best model configuration from the neural architecture search. To obtain classifications of remission, the model's regression outputs were thresholded post-hoc using the clinical criteria for MDD remission (HAMD(week 8) < 7). *RMSE*: root mean squared error, *NNS*: number needed to screen, *PPV*: positive predictive value, *AUC*: area under the receiver operating characteristic curve.

Target	Performance
ΔHAMD	R^2 0.26 (95% CI 0.12–0.40), RMSE 4.45
Remission	NNS 3.2, PPV 0.64, NPV 0.81, AUC 0.71

(HAMD(week 8) $<=$ 7) with number of subjects needed-to-treat (NNT) of 3.2 subjects and AUC of 0.71. This NNT indicates that, on average, one additional remitter will be identified for every 3 individuals screened by this predictor. In comparison, clinically-adopted pharmacological and psychotherapeutic treatments for MDD have NNTs ranging from 2–25 [18], and other proposed predictors for antidepressants besides bupropion have reported NNTs of 3–5 [9,10]. Therefore, this NNT of 3.2 has high potential for clinical benefit in identifying individuals mostly likely to respond to bupropion (Table 2).

When evaluated on sertraline and placebo-treated subjects from the this dataset, the model demonstrated poor accuracy (negative R^2), which is desirable because it indicates the model learned features specific to bupropion response. Additionally, clinical covariates such as demographics, disease duration, and baseline clinical scores were added to the data in another NAS, but this did not increase predictive power. Lastly, less statistically complex models, including multiple linear regression and a support vector machine, performed poorly with negative R^2, even after hyperparameter optimization with a comparable random search of 800 configurations. This finding suggests that a model with a higher statistical capability such as a neural network was needed to learn the association between the data and treatment outcome.

3.2 Learned Neuroimaging Biomarker

Permutation feature importance was measured on the best-performing model configuration to extract a composite neuroimaging biomarker of bupropion response. Specifically, for each feature, the change in R^2 was measured after randomly permuting the feature's values among the subjects. This was repeated 100 times per feature, and the mean change in R^2 provided an estimate of the importance of each feature in accurate predicting bupropion response. The 10 most important regions for bupropion response prediction are visualized in Fig. 3 and include the medial frontal cortex, amygdala, cingulate cortex, and striatum. The regions this model has learned to use agree with the regions neurobiologists have identified as key regions in the reward processing neural circuitry [4]. This circuit is the putative target of bupropion and the circuit largely measured by the reward expectancy task in this task-based fMRI study.

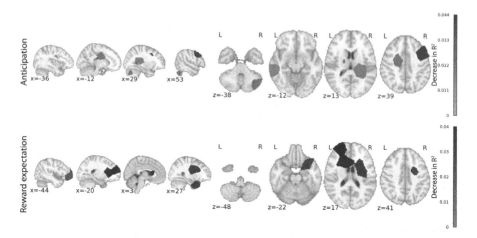

Fig. 3. The 10 most important ROIs for bupropion response prediction, as measured by permutation feature importance. These included 5 regions in the anticipation contrast map (C_{antic}, top row) and 5 regions in the reward expectation contrast map (C_{re}, bottom row). Darker hues indicate greater importance in predicting ΔHAMD.

4   Conclusions

In this work, deep learning and an extensive, unbiased NAS were used to construct predictors of bupropion response from pretreatment task-based fMRI. These methods produced a novel, accurate predictive tool to screen for MDD patients likely to respond to bupropion, to estimate the degree of long-term symptomatic improvement after treatment, and to identify patients who will not respond appreciably to the antidepressant. Predictors such as the one presented are an important step to help narrow down the set of candidate antidepressants to be tested for each patient and to address the urgent need for individualized treatment planning in MDD. The results presented also underscore the value of fMRI and in MDD treatment prediction, and future work will target extension to additional treatments.

References

1. Bergstra, J., Bengio, Y.: Random search for hyper-parameter optimization. J. M. L. Res. **13**, 281–305 (2012)
2. Calhoun, V.D., et al.: The impact of T1 versus epi spatial normalization templates for fMRI data analyses. Hum. Brain Mapp. **38**, 5331–5342 (2017). https://doi.org/10.1002/hbm.23737
3. Chan, M.Y., et al.: Socioeconomic status moderates age-related differences in the brain's functional network organization and anatomy across the adult lifespan. PNAS **115**, E5144–E5153 (2018). https://doi.org/10.1073/pnas.1714021115
4. Chau, D.T., et al.: The neural circuitry of reward and its relevance to psychiatric disorders. Curr. Psych. Rep. **6**(5), 391–399 (2004). https://doi.org/10.1007/s11920-004-0026-8

5. Craddock, R.C., et al.: A whole brain fMRI atlas generated via spatially constrained spectral clustering. Hum. Brain Mapp. **33**(8), 1914–1928 (2012). https://doi.org/10.1002/hbm.21333

6. Dohmatob, E., et al.: Inter-subject registration of functional images: do we need anatomical images? Front. Neurosci. **12**, 64 (2018). https://doi.org/10.3389/fnins.2018.00064

7. Dupuy, J.M., et al.: A critical review of pharmacotherapy for major depressive disorder. Int. J. Neuropsychopharmacol. **14**(10), 1417–1431 (2011). https://doi.org/10.1017/S1461145711000083

8. Etkin, A., et al.: Resolving emotional conflict: a role for the rostral anterior cingulate cortex in modulating activity in the amygdala. Neuron **51**(6), 871–882 (2006). https://doi.org/10.1016/j.neuron.2006.07.029

9. Etkin, A., et al.: A cognitive-emotional biomarker for predicting remission with antidepressant medications: a report from the iSPOT-D trial. Neuropsychopharmacol. Off. Publ. Am. Coll. Neuropsychopharmacol. **40**(6), 1332–1342 (2015). https://doi.org/10.1038/npp.2014.333

10. Gordon, E., et al.: Toward an online cognitive and emotional battery to predict treatment remission in depression. Neuropsychiatr. Dis. Treat. **11**, 517–531 (2015). https://doi.org/10.2147/NDT.S75975

11. Greenberg, P.E., et al.: The economic burden of adults with major depressive disorder in the United States (2005 and 2010). J. Clin. Psych. **76**(2), 155–162 (2015). https://doi.org/10.4088/JCP.14m09298

12. Kessler, R.C., Bromet, E.J.: The epidemiology of depression across cultures. Ann. Rev. Public Health **34**, 119–138 (2013). https://doi.org/10.1146/annurev-publhealth-031912-114409

13. Lener, M.S., Iosifescu, D.V.: In pursuit of neuroimaging biomarkers to guide treatment selection in major depressive disorder: a review of the literature. Ann. NYAS **1344**, 50–65 (2015). https://doi.org/10.1111/nyas.12759

14. Li, L., Talwalkar, A.: Random search and reproducibility for neural architecture search (2019). arXiv:1902.07638

15. Patel, K., et al.: Bupropion: a systematic review and meta-analysis of effectiveness as an antidepressant. Ther. Adv. Psychopharm. **6**, 99–144 (2016). https://doi.org/10.1177/2045125316629071

16. Phillips, M.L., et al.: Identifying predictors, moderators, and mediators of antidepressant response in major depressive disorder: neuroimaging approaches. AJP **172**(2), 124–138 (2015). https://doi.org/10.1176/appi.ajp.2014.14010076

17. Pizzagalli, D.A.: Frontocingulate dysfunction in depression: toward biomarkers of treatment response. Neuropsychopharmacology **36**(1), 183–206 (2011). https://doi.org/10.1038/npp.2010.166

18. Roose, S.P., et al.: Practising evidence-based medicine in an era of high placebo response: number needed to treat reconsidered. Brit. J. Psych. **208**(5), 416–420 (2016). https://doi.org/10.1192/bjp.bp.115.163261

19. Rush, A.J., et al.: Acute and longer-term outcomes in depressed outpatients requiring one or several treatment steps: a STAR* D report. AJP **163**(11), 1905–1917 (2006). https://doi.org/10.1176/ajp.2006.163.11.1905

20. Schaefer, A., et al.: Local-global parcellation of the human cerebral cortex from intrinsic functional connectivity MRI. Cereb. Cortex **28**(9), 3095–3114 (2018). https://doi.org/10.1093/cercor/bhx179

21. Somerville, L.H., et al.: Interactions between transient and sustained neural signals support the generation and regulation of anxious emotion. Cereb. Cortex **23**, 49–60 (2012). https://doi.org/10.1093/cercor/bhr373

22. Trivedi, M.H., et al.: Evaluation of outcomes with citalopram for depression using measurement-based care in STAR* D: implications for clinical practice. AJP **163**(1), 28–40 (2006). https://doi.org/10.1176/appi.ajp.163.1.28
23. Trivedi, M.H., et al.: Establishing moderators and biosignatures of antidepressant response in clinical care (EMBARC): rationale and design. J. Psych. Res. **78**, 11–23 (2016). https://doi.org/10.1016/j.jpsychires.2016.03.001
24. Varma, S., Simon, R.: Bias in error estimation when using cross-validation for model selection. BMC Bioinform. **7**, 91–99 (2006). https://doi.org/10.1186/1471-2105-7-91

Progressive Infant Brain Connectivity Evolution Prediction from Neonatal MRI Using Bidirectionally Supervised Sample Selection

Olfa Ghribi[1,2], Gang Li[3], Weili Lin[3], Dinggang Shen[3], and Islem Rekik[2(✉)]

[1] BASIRA Lab, Faculty of Computer and Informatics,
Istanbul Technical University, Istanbul, Turkey
[2] National School of Engineering of Sfax, University of Sfax, Sfax, Tunisia
irekik@itu.edu.tr
[3] Department of Radiology and BRIC, University of North Carolina at Chapel Hill,
Chapel Hill, NC, USA
http://basira-lab.com

Abstract. The early postnatal developmental period is highly dynamic, where brain connections undergo both growth and pruning processes. Understanding typical brain connectivity evolution would enable us to spot abnormal connectional development patterns. However, this generally requires the acquisition of longitudinal neuroimaging datasets that densely cover the first years of postnatal development. This might not be easily investigated since neonatal follow-up scans are rarely acquired in a clinical setting. Furthermore, waiting for the acquisition of later brain scans would hinder early neurodevelopmental disorder diagnosis. To solve this problem, we unprecedentedly propose a bidirectionally supervised sample selection framework, while leveraging the time-dependency between consecutive observations, for predicting neonatal brain connectome evolution from a single structural magnetic resonance imaging (MRI) acquired around birth. Specifically, we propose to learn how to select the best training samples by *supervisedly* training a *bidirectional ensemble of regressors* from the space of pairwise neonatal connectome disparities to their *expected* prediction scores resulting from using one training connectome to predict another training connectome. The proposed supervised ensemble learning is time-dependent and has a recall memory anchored at the ground truth baseline observation, allowing to progressively pass over previous predictions through the connectome evolution trajectory. We then rank training samples at current timepoint t_{i-1} based on their expected prediction scores by the ensemble and average their connectomes at follow-up timepoint t_i to predict the testing connectome at t_i. Our framework significantly outperformed comparison methods in leave-one-out cross-validation.

© Springer Nature Switzerland AG 2019
I. Rekik et al. (Eds.): PRIME 2019, LNCS 11843, pp. 63–72, 2019.
https://doi.org/10.1007/978-3-030-32281-6_7

1   Introduction

Our brain is a complex network of connectivities, which can capture the functional, structural, and morphological relationships between different brain regions of interest (ROIs). Magnetic resonance imaging (MRI) allows to non-invasively examine the developmental aspects of postnatal brain connectivity in the most dynamic period of brain development, e.g., from birth till the first year of postnatal development. Typically, a brain network, namely a connectome, is modeled as a graph in which each node denotes an anatomical ROI in the brain. The weight between two nodes encodes the relationship between two ROIs (e.g., similarity in morphology or neural activity). Broadly, the estimation of postnatal brain developmental trajectories offers unprecedented opportunities to detect neurodevelopmental disorders such as autism in an early stage [1], estimate dynamic maps of cortical morphology [2], and predict the age of infant brain [3]. However, such rich and insightful studies depend on the availability of a longitudinal dataset, ideally with dense and complete acquisition timepoints in the first year. Up to this point, neuroimaging neonatal and postnatal *longitudinal* datasets remain scarce with the exception of the recent Baby Connectome Project (BCP) [4], which extends the Human Connectome Project from birth through early childhood, providing a unique longitudinal MRI dataset to map brain connectivity developmental trajectories. To circumvent the scarcity of spatiotemporal neonatal and postnatal MRI data, one can leverage predictive machine learning techniques to learn how to accurately *foresee* the developmental trajectory of the baby brain connectome from a single neonatal timepoint. Such models would incite the proliferation of connectomic longitudinal studies by learning how to estimate follow-up missing observations from a single timepoint with minimal neuroimaging resources. Interestingly, to the best of our knowledge, such models of infant brain connectome evolution prediction are absent.

There has been a few works [2,5–7] on predicting the developmental trajectory of the baby brain, mainly focusing on predicting the development of cortical surface shapes [5], cortical attribute maps [2], white matter diffusion fibers [6], and T1-w images from a single MRI acquisition timepoint around birth [7]. The majority of these works were rooted in the following assumption: *if one can learn how to identify the best representative training samples for a given testing subject at baseline timepoint, one can use a weighted average of their corresponding follow-up training samples to predict the missing follow-up data.* Basically, the selected training samples act as proxies linking the neonatal observation to the follow-up observations for predicting the postnatal evolution trajectory of a neonatal brain scan acquired around birth timepoint t_0. To relax this strong assumption, a few works [5,6] integrated an individualization step, where the testing sample is used to locally morph the selected neonatal training samples. Next, by applying those individualizing sample morphing to the follow-up samples at acquisition timepoints, the target developmental trajectory is predicted. Although compelling, the designed sample selection strategies in the prediction step in all those methods have two major limitations. The *first* one lies in overlooking the time-dependency between two consecutive observations at timepoints t_i and t_{i+1}. In fact, all follow-up timepoints are *jointly* predicted from baseline data observed at t_0. The *second* limitation lies in unsupervisedly selecting the best neonatal training samples for the target prediction tasks, while solely relying on their similarity to the neonatal testing subject. Hence, the selected training samples would not necessarily boost the prediction results.

To address these limitations, we propose a bidirectionally supervised sample selection method for predicting baby connectome evolution from a single timepoint in a progressive manner, where the predicted connectome of a testing infant at t_i serves as a guidance along with the baseline testing connectome to predict the T follow-up connectomes at $\{t_i\}_{i=1}^{T}$. In particular, we aim to predict the brain connectivity development from the baseline to the follow-up timepoints *progressively*. Besides, given a neonatal testing connectome at t_0, we propose a novel bidirectional training sample selection strategy, supervised by the *expected* prediction score of selecting a particular baseline training sample to predict the missing follow-up connectome trajectory.

The contributions of our paper span multiple directions and present different kinds of advances:

- *Conceptual advance.* To the best of our knowledge, this is the first work that aims to learn how to predict the developmental trajectory of a neonatal brain connectome over the first year of postnatal brain development.
- *Technical advance.* We propose a novel technique for learning how to select the best neighboring training connectomes for a given testing connectome at baseline. We specifically train *bidirectional ensemble of regressors* to learn how to identify the training samples in a *supervised* manner, which if selected for predicting the neonatal testing connectome evolution trajectory, would yield to the best prediction accuracy for the testing neonate. We also leverage both neonatal testing connectome (i.e., ground truth) along with the predicted connectome at a particular timepoint to jointly predict the follow-up connectome in a *progressive* manner.
- *Generic methodological advance.* The proposed core method is generic and can be evaluated on any anatomical region of interest including the whole brain. It can be also utilized to improve the prediction accuracy by existing image and shape-based evolution prediction frameworks [5–7], which typically use simple strategies for training sample selection.

2   Bidirectionally Supervised Progressive Brain Connectome Evolution Trajectory Prediction

In this section, we detail the key steps of the proposed bidirectionally supervised selection strategy for progressively predicting brain connectome evolution trajectories over different MRI acquisition timepoints. Figure 1 illustrates the key steps of the proposed connectome-specific evolution trajectory prediction framework over time using the designed supervised learning strategy. This strategy charts the selection of the best training brain connectome samples for the target prediction task. Using the proposed supervised sample selection strategy, we first learn a mapping function f_{t_i} at each timepoint t_i, which aggregates two bidirectional regressors $f_{t_i}^+$ and $f_{t_i}^-$, to map the dissimilarity vector between a pair of connectomes to a prediction score. To train both models, we define a prediction score for each pair of training connectomes as a combination of two prediction scores generated when (i) using the first training connectome

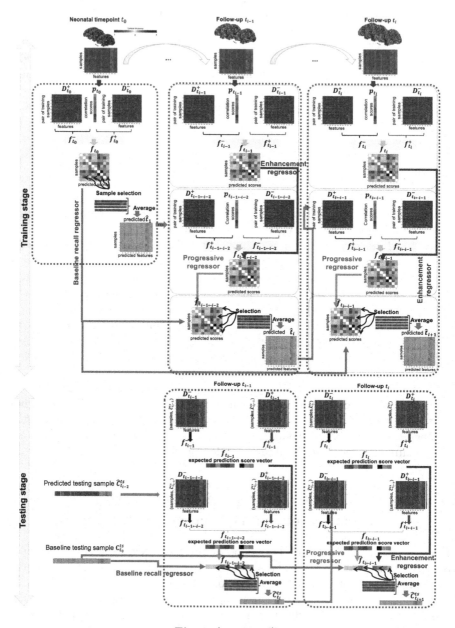

Fig. 1. (continued)

◄**Fig. 1.** (continued) *Pipeline of the proposed bidirectionally progressive prediction frame-work of brain connectome evolution trajectory from neonatal connectome* $\mathbf{C}_{t_0}$. In the training stage, we design and train an ensemble regressor $\bar{f}_{t_{i-1}}$, where each regressor learns a directional mapping ($+$ or $-$ direction) from a disparity matrix $\mathbf{D}_{t_{i-1}}$, encod-ing the pairwise differences between training connectomes, to the expected prediction score vector $\mathbf{p}_{t_{i-1}}$ at each timepoint t_{i-1}. Particularly, we define (i) a progressive regres-sive $f_{t_{i-1} \leftarrow t_{i-2}}$ capturing the time-dependency between consecutive observations, (ii) an enhancement regressor f_{t_i} enhancing the target prediction task at timepoint t_{i-1}, and (iii) a baseline recall regressor f_{t_0} which is passed over to all follow-up trained regressors, acting as a recall memory. Given a testing neonatal connectome $\mathbf{C}_{t_0}^{tst}$, we progressively predict its follow-up connectomes by testing $\bar{f}_{t_{i-1}}$ at each timepoint t_{i-1}. Specifically, we select the training samples at t_{i-1} with the highest expected predic-tion scores outputted by $\bar{f}_{t_{i-1}}$, then average their corresponding observations at t_i to predict $\hat{\mathbf{C}}_{t_i}^{tst}$.

to predict the evolution trajectory of the second training connectome, and (ii) using the second training connectome to predict the evolution trajectory of the first training connectome. This sample selection strategy is supervised by the *expected* prediction performance of the selected training samples in generating authentic connectomes at follow-up timepoints $\{t_i\}_{i=1}^T$. Given a testing neonatal connectome at t_0, we use the trained bidirectional regressors to predict a pair of prediction scores for all training samples at each follow-up timepoint t_i. This allows to select the best training samples for predicting the evolution trajectory of the testing connectome. Last, we linearly average the follow-up connectomes at $\{t_i\}_{i=1}^T$ of the selected training samples to predict the testing connectome at each timepoint t_i. We detail below the contributions of the proposed strategy.

A. Bidirectionally Supervised Sample Selection (Training Stage). Each observed sample s at timepoint t_i is represented by a brain connectome $\mathbf{C}_{t_i}^s$ encoded in a symmetric matrix, where each element quantifies the connectivity strength between a pair of ROIs. We used leave-one-out cross-validation (LOO-CV) to train the proposed supervised brain connectome selection framework aiming to learn how to maximize the *expected* prediction score for a testing con-nectome evolution trajectory. Specifically, to predict a testing brain connectome $\mathbf{C}_{t_i}^{tst}$ for testing sample tst at a specific timepoint t_i, we learn how to identify the best training samples s and s' that maximize the prediction score of $\mathbf{C}_{t_i}^{tst}$ at timepoint t_i, $i \geq 1$. Specifically, given $n-1$ training samples, we train our super-vised sample selection model for each left-out testing connectome $\mathbf{C}_{t_i}^{tst}$. To do so, for each pair of training samples $(\mathbf{C}_{t_i}^s, \mathbf{C}_{t_i}^{s'})$ in the training set comprising $n-1$ samples, we compute their element-wise difference. Excluding self-differences, this generates a *disparity matrix* with $((n-1) \times (n-2))$ rows at each time point t_i, where each row $\mathbf{D}(s, s') = |\mathbf{C}_{t_i}^s - \mathbf{C}_{t_i}^{s'}|$ represents an element-wise difference between two training samples $\mathbf{C}_{t_i}^s$ and $\mathbf{C}_{t_i}^{s'}$.

Next, we define a prediction score $\mathbf{p}_{t_i}^{s,s'}$ as the element (s, s') of vector $\mathbf{p}_{t_i} \in \mathbb{R}^{(n-1) \times (n-2)}$, quantifying the similarity between a training sample $\mathbf{C}_{t_i}^{s'}$

and a target $\mathbf{C}_{t_i}^s$ at t_i. Specifically, we use Pearson correlation (PC) between both samples. We then learn a support vector regressor (SVR) function f_{t_i} that maps the disparity vector $\mathbf{D}(s, s')$ of each pair of training samples $(\mathbf{C}_{t_i}^s, \mathbf{C}_{t_i}^{s'})$, onto their corresponding score value $\mathbf{p}_{t_i}^{s,s'}$. To take into account the *directionality* of the disparity vector since $\mathbf{C}_{t_i}^{s'} - \mathbf{C}_{t_i}^s \neq \mathbf{C}_{t_i}^s - \mathbf{C}_{t_i}^{s'}$, we design two disparity matrices and train bidirectional regressors for our prediction task. Essentially, each row in the 'positive' unidirectional disparity matrix represents the disparity vector between two training samples $\mathbf{C}_{t_i}^s$ and $\mathbf{C}_{t_i}^{s'}$. It is defined as $\mathbf{D}_{t_i}^+(s, s') = max(0, \mathbf{C}_{t_i}^s - \mathbf{C}_{t_i}^{s'})$ (resp. $\mathbf{D}_{t_i}^-(s, s') = max(0, \mathbf{C}_{t_i}^{s'} - \mathbf{C}_{t_i}^s)$ in the 'negative' unidirectional disparity matrix). Next, we train two separate regressors: one regressor function $f_{t_i}^+$ on the positive disparity matrix and another regressor function $f_{t_i}^-$ on the negative disparity matrix. Each regressor function learns a mapping from its input disparity matrix to the target prediction vector $\mathbf{p}_{t_i}$ at timepoint t_i. To aggregate the expected prediction scores by both trained regressors, we further define the final weighted f_{t_i} regressor at timepoint t_i as: $f_{t_i} = \mu_{t_i} \times exp(|f_{t_i}^+ - f_{t_i}^-|)$, where the weight $\mu_{t_i} = (f_{t_i}^+ + f_{t_i}^-)/2$.

B. Progressively Time-Dependent Supervised Sample Selection (Training Stage). We note that in the previous step follow-up connectomes are predicted from the baseline connectome at each timepoint *independently*. However, the predicted connectome at timepoint t_{i-1} can be leveraged to guide the prediction at follow-up timepoint t_i. As such, the connectome evolution trajectory is learned in a progressive way while assimilating the time-dependency between consecutive observations. Similarly, we learn the pairwise similarity between predicted training samples at t_{i-1} and the ground-truth training samples at timepoint t_i. Therefore, we train a positive *progressive* regressor $f_{t_i \leftarrow t_{i-1}}^+$ at timepoint t_i between two training samples s and s' dependently on the previous timepoint t_{i-1}. Specifically, we define it as: $f_{t_i \leftarrow t_{i-1}}^+(\mathbf{D}_{t_i \leftarrow t_{i-1}}^+(s, s'))$, where $\mathbf{D}_{t_i \leftarrow t_{i-1}}^+(s, s') = max(0, \mathbf{C}_{t_i}^s - \mathbf{C}_{t_{i-1}}^{s'})$. Similarly, we train a negative *progressive* regressor $f_{t_i \leftarrow t_{i-1}}^-$ on the negative disparity matrix. Specifically, we define it as: $f_{t_i \leftarrow t_{i-1}}^-(\mathbf{D}_{t_i \leftarrow t_{i-1}}^-(s, s'))$, where $\mathbf{D}_{t_i \leftarrow t_{i-1}}^-(s, s') = max(0, \mathbf{C}_{t_{i-1}}^{s'} - \mathbf{C}_{t_i}^s)$. The positive (resp. negative) regressor function maps each disparity vector $\mathbf{D}_{t_i \leftarrow t_{i-1}}^+(s, s')$ (resp. $\mathbf{D}_{t_i \leftarrow t_{i-1}}^-(s, s')$) to a target expected prediction score $\mathbf{p}_{t_i \leftarrow t_{i-1}}^{s,s'}$ defined as the Pearson correlation between training connectomes $\mathbf{C}_{t_i}^s$ and $\mathbf{C}_{t_{i-1}}^{s'}$. In the spirit of the previous step, we further propose a weighted progressive regressor function $f_{t_i \leftarrow t_{i-1}}$ aggregating both positive and negative progressive regressors as follows:

$$\begin{cases} f_{t_i \leftarrow t_{i-1}} = \mu_{t_i \leftarrow t_{i-1}} \times exp(|f_{t_i \leftarrow t_{i-1}}^+ - f_{t_i \leftarrow t_{i-1}}^-|) \\ \mu_{t_i \leftarrow t_{i-1}} = (f_{t_i \leftarrow t_{i-1}}^+ + f_{t_i \leftarrow t_{i-1}}^-)/2 \end{cases}$$

More importantly, instead of solely using $f_{t_i \leftarrow t_{i-1}}$ to aggregate the expected prediction scores by both regressors for each training sample in the selection step, we further propose an ensemble regressor $\bar{f}_{t_i}$, where we pass over through all follow-up timepoints the estimated score by f_{t_0} and estimate a final regressor

$\bar{f}_{t_i}$ with a memory anchored at baseline f_{t_0} (baseline recall regressor). This plays an important role in the testing stage as it enables to track local connectomic changes based on the only ground truth observation we have: $\mathbf{C}_{t_0}^{tst}$. We further integrate the regressor f_{t_i} (enhancement regressor) at prediction timepoint t_i to enhance the target prediction task at timepoint t_i. We formulate this new type of ensemble regressor as follows:

$$\bar{f}_{t_i} = (\underbrace{f_{t_i}}_{\text{enhancement regressor}} + \underbrace{f_{t_i \leftarrow t_{i-1}}}_{\text{progressive regressor}} + \underbrace{f_{t_0}}_{\text{baseline recall regressor}})/3$$

C. Progressive Neonatal Connectome Evolution Trajectory Prediction (Testing Step). In the testing stage, given a testing neonatal connectome $\mathbf{C}_{t_0}^{tst}$, we first compute the bidirectional disparity vectors at baseline timepoint t_0 between $\mathbf{C}_{t_0}^{tst}$ and each connectome $\mathbf{C}_{t_0}^{s}$ of training sample s. Next, we predict the weight score for each training sample using f_{t_0} defined in step (A) since the baseline timepoint does not have a previous observation. We then sort the training samples by their expected prediction scores by f_{t_0} in decreasing order and select the top K with the highest scores. By averaging their corresponding K connectomes at timepoint t_1, we predict $\hat{\mathbf{C}}_{t_1}^{tst}$. Follow-up connectomes at $\{t_i\}_{i=2}^{T}$ are then predicted in a *progressive* manner. Basically, given the predicted testing connectome at timepoint t_{i-1}, we predict its evolution at t_i as follows. First, we compute the pairwise positive distance $\mathbf{D}_{t_i \leftarrow t_{i-1}}^{+}(\mathbf{C}_{t_i}^{s}, \hat{\mathbf{C}}_{t_{i-1}}^{tst})$ between each training sample $\mathbf{C}_{t_i}^{s}$ and the predicted testing brain connectome $\hat{\mathbf{C}}_{t_{i-1}}^{tst}$. Second, we predict the similarity score $\mathbf{p}_{t_i}^{s,tst}$ between them by testing the learned regression function $f_{t_i \leftarrow t_{i-1}}^{+}$. In a similar way, we predict the second similarity score using their negative distance vector $\mathbf{D}_{t_i \leftarrow t_{i-1}}^{-}(\mathbf{C}_{t_i}^{s}, \hat{\mathbf{C}}_{t_{i-1}}^{tst})$ and the learned regression function $f_{t_i \leftarrow t_{i-1}}^{-}$. Third, we use the ensemble regressor $\bar{f}_{t_i}$ trained in the previous step (B) to predict the score associated with each training sample for our target prediction task. Ultimately, by sorting all training samples at each follow-up timepoint according to the outputted scores by $\bar{f}_{t_i}$, we predict the testing sample $\hat{\mathbf{C}}_{t_i}^{tst}$ at a timepoint t_i by averaging the top K selected training samples at t_i with the highest prediction scores.

3 Results and Discussion

Dataset and Parameters. We evaluated the proposed framework on 11 typically developing infants, each with 4 longitudinal T1-w and T2-w MRI scans acquired at 1, 3, 6, and 9 months of age. After rigid alignment of longitudinal and cross-sectional infant structural MR images (i.e., T1-w and T2-w) and brain tissue segmentation, we reconstructed and parcellated each cortical surface of each hemisphere into 35 regions using Desikan-Killiany cortical parcellation proposed in [8]. By computing the pairwise absolute difference in cortical thickness between pairs of regions, we generate a 35×35 morphological connectivity matrix (i.e., cortical morphological connectome) for each time point of each subject as in [9,10]. We set the number of selected neighbors to $K = 4$ across all methods for fair comparison.

Evaluation. Since there is no existing work on brain connectome evolution prediction from baseline, we compared our bidirectionally supervised progressive prediction (BSPP) method against its three variants. **(1) Bidirectionally supervised prediction (BSP):** is based only on step (A) without progressive prediction, where the best training samples are selected at each timepoint *independently* in a supervised manner. **(2) Bidirectionally supervised joint prediction (BSJP):** is based only on step (A) without progressive prediction, where the best training samples are selected *solely* at baseline timepoint t_0 to jointly predict all follow-up timepoints. **(3) Joint prediction (JP):** is based on selecting the most similar baseline training samples to the baseline testing sample using Pearson correlation as a similarity measure. Follow-up timepoints are then predicted *jointly* based on the selected samples as in [5].

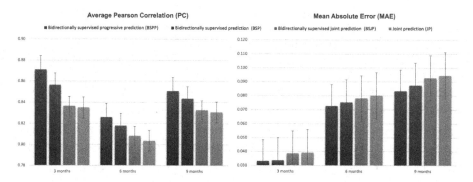

Fig. 2. Prediction results using average pearson correlation coefficient (PC) and mean absolute error (MAE) by the proposed and comparison methods.

Besides, given the small size of our dataset, we used LOO-CV to train our proposed method and its variants. To evaluate the performance of each method, we computed the average Pearson correlation coefficient between each predicted brain connectome and its ground truth across all subjects at each follow-up timepoint. The higher the Pearson correlation coefficient the more similar is the predicted brain connectome to the ground truth. Additionally, we computed the mean average error (MAE) between each predicted connectome and its ground truth. The lower the MAE score the more reliable is our prediction. Figure 2 displays the prediction results by all methods. Remarkably, the proposed BSPP achieves the best prediction results in comparison with its variants ($p - value < 0.05$ using two-tailed paired t-test). It is worth noting that when using Pearson correlation, the prediction accuracy drops at 6 months for all methods. This can be explained by low MRI tissue contrast between gray and white matter tissues at 6 months of age, which might produce a noisy reconstruction of the cortical surfaces around this age. This might influence the construction of the cortical morphological connectome across individuals. Using MAE, we notice that the prediction error increases over time (from 3 to 9 months). This might be explained by the *dynamic* cortical thickness increase during cortical development.

The proposed BSPP method has the advantage to explore the relationship between samples across various timepoints. This permits to progressively update the relationship modeled in the learning of the ensemble regressor. Hence, the best training samples selected at each timepoint might be different. This contribution enables our model to well capture the dynamic connectome developmental trajectory compared to joint prediction methods. However, the performance of this supervised method depends on the presence of training samples which are sufficiently similar to the testing connectome. Since we inherently assume that the set of predictors of the closest neighbors to the testing sample are also good predictors of the testing subject, using a larger longitudinal dataset would demonstrate the scalability of our method. The proposed BSPP can also be leveraged to predict *atypical* neurodevelopmental connectome evolution as in the recent works [11,12] on predicting the evolution of disordered brain T1-weighted MRI from baseline.

4   Conclusion

This paper focuses on a new problem on baby connectome evolution trajectory prediction from a single neonatal observation. To achieve this, we propose a novel bidirectionally supervised progressive prediction model which is based on aggregating different types of regressors for learning how to effectively select the best training samples with the highest expected prediction scores for the target testing sample. Experiments show that our method can significantly boost the prediction accuracy at all follow-up timepoints. This paper opens a new window for baby connectome evolution prediction. There are many possible future directions yet to explore, such as developing multimodal trajectory predictor using functional and structural connectomes.

References

1. Hazlett, H.C., et al.: Early brain development in infants at high risk for autism spectrum disorder. Nature **542**, 348 (2017)
2. Meng, Y., Li, G., Gao, Y., Lin, W., Shen, D.: Learning-based subject-specific estimation of dynamic maps of cortical morphology at missing time points in longitudinal infant studies. Hum. Brain Mapp. **37**, 4129–4147 (2016)
3. Dean III, D.C., et al.: Estimating the age of healthy infants from quantitative myelin water fraction maps. Hum. Brain Mapp. **36**, 1233–1244 (2015)
4. Howell, B.R., et al.: The UNC/UMN baby connectome project (BCP): an overview of the study design and protocol development. NeuroImage **185**, 891–905 (2019)

5. Rekik, I., Li, G., Lin, W., Shen, D.: Predicting infant cortical surface development using a 4D varifold-based learning framework and local topography-based shape morphing. Med. Image Anal. **28**, 1–12 (2016)
6. Rekik, I., Li, G., Yap, P.T., Chen, G., Lin, W., Shen, D.: Joint prediction of longitudinal development of cortical surfaces and white matter fibers from neonatal MRI. NeuroImage **152**, 411–424 (2017)
7. Rekik, I., Li, G., Wu, G., Lin, W., Shen, D.: Prediction of infant MRI appearance and anatomical structure evolution using sparse patch-based metamorphosis learning framework. In: Wu, G., Coupé, P., Zhan, Y., Munsell, B., Rueckert, D. (eds.) Patch-MI 2015. LNCS, vol. 9467, pp. 197–204. Springer, Cham (2015). https://doi.org/10.1007/978-3-319-28194-0_24
8. Li, G., Wang, L., Shi, F., Lin, W., et al.: Simultaneous and consistent labeling of longitudinal dynamic developing cortical surfaces in infants. Med. Image Anal. **18**, 1274–1289 (2014)
9. Rekik, I., Li, G., Lin, W., Shen, D.: Estimation of brain network atlases using diffusive-shrinking graphs: application to developing brains. In: Niethammer, M., et al. (eds.) IPMI 2017. LNCS, vol. 10265, pp. 385–397. Springer, Cham (2017). https://doi.org/10.1007/978-3-319-59050-9_31
10. Dhifallah, S., Rekik, I., Alzheimer's Disease Neuroimaging Initiative and others: Clustering-based multi-view network fusion for estimating brain network atlases of healthy and disordered populations. J. Neurosci. Methods **311**, 426–435 (2019)
11. Gafuroğlu, C., Rekik, I., Authorinst for the Alzheimer's Disease Neuroimaging Initiative: Joint Prediction and Classification of Brain Image Evolution Trajectories from Baseline Brain Image with Application to Early Dementia. In: Frangi, A.F., Schnabel, J.A., Davatzikos, C., Alberola-López, C., Fichtinger, G. (eds.) MICCAI 2018. LNCS, vol. 11072, pp. 437–445. Springer, Cham (2018). https://doi.org/10.1007/978-3-030-00931-1_50
12. Gafuroğlu, C., Rekik, I.: Image evolution trajectory prediction and classification from baseline using learning-based patch atlas selection for early diagnosis. arXiv preprint arXiv:1907.06064 (2019)

Computed Tomography Image-Based Deep Survival Regression for Metastatic Colorectal Cancer Using a Non-proportional Hazards Model

Alexander Katzmann[1,4]($\boxtimes$) (iD), Alexander Mühlberg[1], Michael Sühling[1],
Dominik Nörenberg[2], Stefan Maurus[2], Julian Walter Holch[3],
Volker Heinemann[3], and Horst-Michael Groß[4]

[1] Siemens Healthcare GmbH, Computed Tomography, 91301 Forchheim, Germany
alexander.katzmann@siemens-healthineers.com
[2] Department of Radiology, University Hospital Großhadern,
Ludwig-Maximilians-University Munich, Marchioninistrasse 15,
81377 Munich, Germany
[3] Department of Internal Medicine III, Comprehensive Cancer Center,
University Hospital Großhadern, Ludwig-Maximilians-University Munich,
Marchioninistrasse 15, 81377 Munich, Germany
[4] Neuroinformatics and Cognitive Robotics Lab,
Ilmenau, University of Technology, 98693 Ilmenau, Germany

Abstract. With more than 1,800,000 cases and over 862,000 deaths per year, metastatic colorectal cancer is the second leading cause of cancer related deaths in modern societies. The estimated patient survival is one of the main factors for therapy adjustment. While proportional hazard models are a key instrument for survival analysis within the last centuries, the assumption of hazard proportionality might be overly restrictive and their applicability to complex data remains difficult. Especially the integration of image data comes at the cost of a careful pre-selection of hand-crafted features only. With the rise of deep learning, directly differentiable models for survival analysis have been developed. While some inherit the difficulties of the proportionality assumption, others are restricted to scalar data input. Computed-tomography-based survival analysis remains a hardly researched topic at all. We propose a deep model for computed-tomography-based survival analysis providing a hazard probability output representation comparable to Cox regression without relying on the hazard proportionality assumption. The model is evaluated on multiple datasets, including metastatic colorectal cancer computed tomography imaging data, and significantly reduces the average prediction error compared to the Cox proportional hazards model.

Keywords: Survival analysis · Deep learning · Computed tomography · Colorectal cancer

I. Rekik et al. (Eds.): PRIME 2019, LNCS 11843, pp. 73–80, 2019.
https://doi.org/10.1007/978-3-030-32281-6_8

1   Introduction

Metastatic colorectal cancer is the second leading cause of cancer related deaths in modern societies. According to the American Cancer Society, metastatic colorectal cancer is expected to cause more than 51,000 deaths during 2019 in the U.S. alone, and more than 862,000 cases worldwide. The lifetime risk of developing colorectal cancer is about 1 in 22 (4.49%) for men and 1 in 24 (4.15%) for women [2,8]. For treatment planning and treatment evaluation, two important quantitative outcome markers are (a) the overall and (b) the progression- or event-free survival, typically measured in days or months after the date of the first diagnosis (DOFD), respectively the first treatment (DOFT) [3,6]. For treatment response assessment and prediction, these measures are usually approximated using statistical models. This process is called survival analysis. Although there is a variety of possible statistical models for survival analysis [17], the most commonly used models are based on product limit estimators, e.g. Kaplan-Meier curves for group-based survival *description*, and parametric estimators, like Cox proportional hazards (CPH) models, for *quantitative modelling* of the influence of categorical or continuous variables on the survival expectation of single individuals within a cohort. Estimating a specific patient's treatment outcome as a time-dependent cumulative hazard function is beneficial for treatment planning, as it provides quantitative information on the expected course of disease and may therefore allow a better therapy planning, as well as an early therapy adaption. As already mentioned, the CPH model is one of the most widespread variants of survival analysis algorithms. It models a patient specific survival hazard $\lambda(t|X_i)$ with a time-dependent base hazard probability $\lambda_0(t)$ time t for all subjects, and by a covariate-dependent, patient specific exponential factor with covariates X:

$$\lambda(t|X) = \lambda_0(t)\exp(\beta_0 X_0 + ... + \beta_n X_n) = \lambda_0(t)\exp(\beta \cdot X) \tag{1}$$

However, this formulation implies several limitations:

- the base hazard probability λ_0 is shared across all individuals, thus, it is independent of a-priori differences between subjects, e.g. different timepoints of therapy begin after the date of the first diagnosis, etc.
- the covariates $X_0, ..., X_n$ affect the base probability as additive exponential factors, thus, they can have no direct interactions
- covariates X are applied to the base hazard probability as time-independent, proportional exponential factors, meaning they affect the base hazard probability *equally at every timepoint* within the therapy

To overcome these restrictions, within the last years another field has evolved which applies the bagging technique of random forests to survival analysis called Random Survival Forests (RSFs) [12]. RSFs can be applied flexibly to survival data, while providing a reasonably well generalization performance. However, RSFs as well as parametric modelling techniques, like CPH models, require predefined, hand-crafted features, making them less suitable for scenarios with non- or semi-structed data, as well as for explorative data analysis.

Currently, deep neural networks are the standard state-of-the-art machine learning approach with deep neural networks being applied for a wide variety of medical applications, often outperforming classical approaches [9]. Recently, deep learning has been employed for survival regression [13,16], allowing survival models to be used for explorative data analysis, e.g. by utilizing image data. Therefore, no handcrafted features are required, contrary to, e.g., Radiomics [15]. Instead, image features useful for survival regression are learned automatically.

The algorithms from [10,13] provide a well-founded deep learning variant of CPH models, combining their advantages with deep feature learning. However, as they are fundamentally based on the CPH assumption, they are subject to the aforementioned restrictions, especially regarding time dependency and interaction of covariates. The algorithm from [16], following an approach similar to the proposed method, builts up on a loss definition which focusses on the relative event time order only, making the algorithm prone to major differences in the absolute time scale.

In contrast, our approach is built up on a single-shot estimate of the individual's hazard curve, providing an immediate measure of both, the base hazard rate over time, as well as the patient specific hazard rate, while taking into account the event observation order as well as the absolute observation times, as will be demonstrated in Sect. 3.

2   Methods

The most commonly used survival regression method is the CPH model. Its regression is based on an estimate of the individual hazard rate h parameterized by the time t and individual influences z_i for individuals i. It is defined by a time-dependent base hazard probability $h_0(t)$ and an exponential fit of z_i with regression coefficients β:

$$h(t|z_i) = h_0(t) \cdot \exp(z_i\beta) \tag{2}$$

The model is fitted sequentially first using a partial likelihood function $L_i(\beta)$ with event observation times o_i for individuals i:

$$L_i(\beta) = \frac{h(o_i|z_i)}{\sum_{j:o_j \geq o_i} h(o_i|z_j)} = \frac{h_0(o_i)\exp(z_i\beta)}{\sum_{j:o_j \geq o_i} h_0(o_i)\exp(z_j)} = \frac{\exp(z_i\beta)}{\sum_{j:o_j \geq o_i} \exp(z_j\beta)} \tag{3}$$

which can be optimized independently from the base hazard rate h_0. Secondly, h_0 can be estimated using the results for β from Eq. 3.

2.1   Concept

Employing a deep neural network to estimate the hazard function $h(t|z_i)$ allows to infer features using an end-to-end optimization process. Moreover it is possible to learn features from images without time-consuming hand-crafted feature design. Similar to the algorithm from [16], our method uses a fixed amount of

K equally sized discrete timesteps. Each timestep is represented by one neuron of a sigmoidal output. Therefore, the observation timepoints o_i have to be transformed to normalized observation timesteps $y_i = \frac{o_i}{\max o}(K-1)$. In contrast to Eq. 2, we omit the hazard proportionality assumption, as it for some scenarios was shown to result in a poor fit for non-proportional or time dependent hazards, which are often present in real data [20]. As mentioned in Sect. 1, our model is based on an estimate of the stationary hazard probabilities $\hat{h}(z_i)$ over time, where stationary means that hazard probabilities are represented assuming survival until that timestep.

2.2   Loss Function

The model can be trained by minimizing the combined loss function. While L_1 takes care of differences between estimated and observed event times, L_2 takes care of right-censored data by penalizing predicted hazards not observed within the observation period. We optimize model parameters Θ for observation times y_i of patients i:

$$\arg \min_{\Theta} L_1^2 + L_2^2 \tag{4}$$

Given the definition from Sect. 2.1, the estimated stationary probabilities $\overline{h}$ of no hazard for each timestep can be derived as:

$$\overline{h}(z_i) = 1 - \hat{h}(z_i) \tag{5}$$

The estimated *non-stationary* probability $\tilde{h}$ of observing an event within a specific timestep $k : 0 \leq k < K$ can be inferred as:

$$\tilde{h}(k|z_i) = \hat{h}(k|z_i) \cdot \prod_{a:0 \leq a < k} \overline{h}(k|z_i) \tag{6}$$

Using the non-stationary probability, one can calculate the expected event observation timestep $\hat{y}_i : 0 \leq \hat{y}_i \leq K - 1, \hat{y}_i \in \mathbb{R}$. The difference loss δ can be calculated as the binary crossentropy of the normalized expected and observed event observation timesteps:

$$\delta(y_i, \hat{y}_i | z_i, \omega_i) = -\omega_i \left(y_i^* \log\left(\hat{y}_i^*\right) + (1 - y_i^*) \log\left(1 - \hat{y}_i^*\right)\right) \tag{7}$$

with ω_i indicating whether an event was observed for patient i, and y_i^* and $\hat{y}_i^*$ being inferred from y_i and $\hat{y}_i$ via division by $K - 1$. For enforcing concordance, an additional term $\mathcal{C}$ is introduced:

$$\mathcal{C} = \sum_{i:0 \leq i < n-1} \sum_{j:i \leq j < n} \left\{ \begin{array}{ll} (\omega_i \cdot \omega_j) \cdot \frac{max(\hat{y}_i^* - \hat{y}_j^*, 0)}{max(\hat{y}_i^*, \hat{y}_j^*)} & \text{if } y_i^* \leq y_j^* \\ 0 & \text{else} \end{array} \right\} \tag{8}$$

Finally, L_1 can be derived as:

$$L_1 = \sum_{i:0 \leq i < n} \delta(y_i, \hat{y}_i | z_i, \omega_i) \ + \ \mathcal{C} \tag{9}$$

Note that this definition of L_1 covers samples for which the event was actually observed, i.e. $\omega_i = 1$. When no event was observed, the probability of an event between the expected and the observed interval equals the error. The summed probability of observing an event within the observation period $D_i = [\hat{y}_i, y_i]$ can then be determined as:

$$\tilde{h}(D_i|z_i) = \sum_{k:\hat{y} \leq k \leq y_i} \tilde{h}(k|z_i) \tag{10}$$

which allows to derive L_2 as:

$$L_2 = \sum_{i:0 \leq i < n} -(1 - \omega_i) \log(1 - \tilde{h}(D_i|z_i)) \tag{11}$$

3 Evaluation

We evaluated our model on the following datasets:

1. SEER Incidence database - covering cancer incidence data from population-based cancer registries with more than 10 million cases.
2. Rossi's Criminal Recidivism dataset - a widely available benchmark dataset used as a proof-of-concept for datasets with only few observations.
3. mCRC dataset - a retrospectively acquired dataset, providing baseline and followup data for more than 200 patients with metastatic colorectal cancer (mCRC), including computed tomography (CT) images and patient metadata (patient demographics, histology, laboratory values, etc.)

For each dataset, the Cox regression and the deep survival regression approach were directly compared. We measured concordance index (CI), mean (MAE) and median absolute error (MedAE). All results were calculated using 10-fold cross validation. Confidence intervals were derived using bootstrapping until convergence [7].

3.1 SEER Incidence Dataset

The SEER incidence dataset contains more than 10,050,000 datasets with more than 130 variables per patient [1,18]. For the evaluation, the colorectal cancer subset was used, as it is best comparable to our target scenario. The subset contains 554,687 datasets. The network was an 8-layer fully-connected ResNet with $k = 60$ output neurons. The input contained age, sex, year of birth, month and year of diagnosis, race, laterality, and TNM-staging [5]. Survival times were cut-off after 3, respectively 5 years, where 3 years is the mean sojourn time for colorectal cancer [21]. With our method, the median and mean absolute error could be significantly reduced compared to the CPH model, archiving relative improvements of 13.2%, resp. 26.9% for 5-years, and 19.1%, resp. 28.1% for 3-years survival, as shown in Table 1. However, a slight drop in the concordance index of 3.6%, resp. 2.2% could be observed. All differences were highly significant ($p < .001$, two-tailed t-test).

Table 1. SEER dataset results with 95% confidence intervals

Algorithm	CI	MAE	$MedAE$
Cox PH 3-y	**.689** [.689, .690]	11.8 [11.7,11.8]	9.00 [9.00,9.00]
Deep Model 3-y	.653 [.652, .654]	**8.49** [8.46,8.51]	**7.28** [7.25,7.31]
Cox PH 5-y	**.682** [.681, .684]	18.2 [18.1,18.3]	13.6 [13.0,14.0]
Deep Model 5-y	.660 [.659, .660]	**13.3** [13.3,13.4]	**11.8** [11.8,11.9]

Table 2. Rossi Criminal Recidivism dataset results with 95% confidence intervals

Algorithm	CI	MAE	MedAE
Cox PH	**.594** [.539, .646]	28.4 [25.6,31.0]	29.1 [24.0,35.0]
Deep	**.587** [.530, .644]	**20.0** [17.6,22.4]	**17.6** [13.7,24.8]

3.2   Rossi Criminal Recidivism Dataset

The Rossi Criminal Recidivism dataset [19] contains 432 datasets with 9 columns, including observation interval and event indicator. Of these, an event is observed for only 114 cases. The Rossi dataset is thus ideal for evaluating tolerance against incomplete observation data. The network architecture was similar to Sect. 3.1. As seen in Table 2, mean and median absolute error could be reduced significantly by 29.6%, resp. 39.5% with $p < .001$. The concordance index again slightly dropped from .594 to .587, also this difference was not significant ($p = .849$, two-tailed t-test).

3.3   mCRC Dataset

For the metastatic colorectal cancer (mCRC) dataset, baseline and followup scans of target liver lesions with radiologic ground-truth annotations from 80 patients were used to predict patient survival, as lesion shape and texture were shown to be predictive for one-year survival in recent work [4,14]. Example data is shown in Fig. 1. A deep convolutional neural network (DCNN) based on

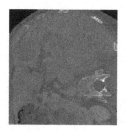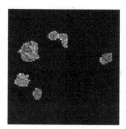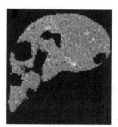

Fig. 1. Example data from the mCRC dataset. Left: liver bounding box of the original CT image; middle/right: masks for selected liver lesions, resp. healthy liver tissue

ResNet [11] was applied. Additionally we injected patient age, tumor grading, RECIST diameters and time after diagnosis into the pre-output layer, which also were the input for the CPH model. While MAE and MedAE again were significantly lower with our approach (see Table 3), the CI indicates that overall both algorithms did not perform well on regressing the overall patient survival from the available data.

Table 3. Results on the mCRC dataset with 95% confidence intervals

Algorithm	CI	MAE	MedAE
Cox PH	**.536** [.508, .569]	423 [397,450]	384 [357,403]
Deep	**.510** [.481, .540]	**325** [301,349]	**262** [233,291]

4   Discussion

We propose a method for deep survival regression being able to handle categorical, scalar, and image data. Although the Cox proportional hazards model generally provides a good measure of concordance between samples, our proposed method helped to significantly reduce mean and median prediction errors on datasets of big, small, and incomplete data. Our method could therefore allow for a better prediction of survival times within clinical contexts, while preserving an inherent measure of hazards over time represented directly within the network architecture. While for the mCRC dataset the mean and median error were reduced significantly, the presumed correlation between target lesions and patient survival could *not* be shown with both tested approaches. We expect this to be explainable by the high data variance compared to the low amount of available data, which is reflected by the small effect sizes reached in a recent publication for one-year survival prediction on the same dataset [14]. The implementation of the proposed method is straight-forward and does not assume a specific problem or network structure, which could make it feasible for a variety of clinical applications, resulting in a need of future work to analyze the concrete circumstances needed for a successful application.

References

1. Surveillance, Epidemiology, and End Results (SEER) program (www.seer.cancer. gov) research data (1973–2015), National Cancer Institute, DCCPS, Surveillance Research Program, released April 2018, based on the November 2017 submission (2017)
2. American Cancer Society: Cancer Facts and Figures. American Cancer Society, Atlanta (2019)
3. National Cancer Institute overall survival (2019). www.cancer.gov/publications/ dictionaries/cancer-terms/def/os. Accessed 01 Apr 2019

4. Aerts, H.J., et al.: Decoding tumour phenotype by noninvasive imaging using a quantitative radiomics approach. Nat. Commun. **5**, 4006 (2014)
5. Brierley, J.D., Gospodarowicz, M.K., Wittekind, C.: TNM Classification of Malignant Tumours. Wiley, Hoboken (2016)
6. Cohen, S.J., et al.: Relationship of circulating tumor cells to tumor response, progression-free survival, and overall survival in patients with metastatic colorectal cancer. Clin. Oncol. **26**, 3213–3221 (2008)
7. Efron, B.: Bootstrap methods: another look at the Jackknife. In: Kotz, S., Johnson, N.L. (eds.) Breakthroughs in Statistics. SSS, pp. 569–593. Springer, New York (1992). https://doi.org/10.1007/978-1-4612-4380-9_41
8. Ferlay, J., et al.: Cancer incidence and mortality worldwide: sources, methods and major patterns in GLOBOCAN 2012. Int. J. Cancer **136**(5), E359–E386 (2015)
9. Greenspan, H., Van Ginneken, B., Summers, R.M.: Guest editorial deep learning in medical imaging: overview and future promise of an exciting new technique. IEEE Trans. Med. Imaging **35**(5), 1153–1159 (2016)
10. Haarburger, C., Weitz, P., Rippel, O., Merhof, D.: Image-based survival analysis for lung cancer patients using CNNs. arXiv preprint arXiv:1808.09679 (2018)
11. He, K., Zhang, X., Ren, S., Sun, J.: Deep residual learning for image recognition. In: Proceedings of the IEEE Conference on Computer Vision and Pattern Recognition, pp. 770–778 (2016)
12. Ishwaran, H., Kogalur, U.B., Blackstone, E.H., Lauer, M.S., et al.: Random survival forests. Ann. Appl. Stat. **2**(3), 841–860 (2008)
13. Katzman, J.L., Shaham, U., Cloninger, A., Bates, J., Jiang, T., Kluger, Y.: Deep survival: a deep cox proportional hazards network. Stat **1050**, 2 (2016)
14. Katzmann, A., et al.: Predicting lesion growth and patient survival in colorectal cancer patients using deep neural networks (2018)
15. Lao, J., et al.: A deep learning-based radiomics model for prediction of survival in Glioblastoma Multiforme. Sci. Rep. **7**(1), 10353 (2017)
16. Lee, C., Zame, W.R., Yoon, J., van der Schaar, M.: DeepHit: a deep learning approach to survival analysis with competing risks. In: Thirty-Second AAAI Conference on Artificial Intelligence (2018)
17. Miller Jr., R.G.: Survival Analysis, vol. 66. Wiley, Hoboken (2011)
18. Ries, L.A.G., et al.: Cancer incidence and survival among children and adolescents: United States SEER Program 1975–1995. Cancer incidence and survival among children and adolescents: United States SEER Program 1975–1995 (1999)
19. Rossi, P.H., Berk, R.A., Lenihan, K.J.: Money, work and crime: some experimental results (1980)
20. Schemper, M.: Cox analysis of survival data with non-proportional hazard functions. J. R. Stat. Soc.: Ser. D (Stat.) **41**(4), 455–465 (1992)
21. Zauber, A.G., et al.: Colonoscopic polypectomy and long-term prevention of colorectal-cancer deaths. New Engl. J. Med. **366**(8), 687–696 (2012)

7 Years of Developing Seed Techniques for Alzheimer's Disease Diagnosis Using Brain Image and Connectivity Data Largely Bypassed Prediction for Prognosis

Mayssa Soussia[1,2] and Islem Rekik[1(✉)]

[1] BASIRA Lab, Faculty of Computer and Informatics,
Istanbul Technical University, Istanbul, Turkey
irekik@itu.edu.tr
[2] Department of Electrical Engineering,
The National Engineering School of Tunis, Tunis, Tunisia
http://basira-lab.com

Abstract. Unveiling pathological brain changes associated with Alzheimer's disease (AD) and its earlier stages including mild cognitive impairment (MCI) is a challenging task especially that patients do not show symptoms of dementia until it is late. Over the past years, neuroimaging techniques paved the way for computer-based diagnosis and prognosis to facilitate the automation of medical decision support and help clinicians identify cognitively intact subjects that are at high-risk of developing AD. As a progressive neurodegenerative disorder, researchers investigated how AD affects the brain using different approaches: (1) *image-based methods* where mainly neuroimaging modalities are used to provide early AD biomarkers, and (2) *network-based methods* which focus on functional and structural brain connectivities to give insights into how AD alters brain wiring. In this *exceptional* review paper, we screened MICCAI proceedings published between 2010 and 2016 and IPMI proceedings published between 2011 and 2017, where 'seed' technical ideas generally get published, to identify neuroimaging-based technical methods developed for AD and MCI classification and prediction tasks. We included papers that fit into image-based or network-based categories. We found out that the majority of papers focused on classifying MCI vs. AD brain states, which has enabled the discovery of discriminative or altered brain regions and connections. However, very few works aimed to *predict* MCI progression from early observations. Despite the high importance of reliably identifying which early MCI patient will convert to AD, remain stable or reverse to normal over months/years, predictive models that *foresee* MCI evolution are still lagging behind.

© Springer Nature Switzerland AG 2019
I. Rekik et al. (Eds.): PRIME 2019, LNCS 11843, pp. 81–93, 2019.
https://doi.org/10.1007/978-3-030-32281-6_9

1   Introduction

[1] Alzheimer's disease (AD), which is the most common form of dementia, is still today an incurable degenerative disease. AD is also known as an irreversible, progressive disorder that destroys neurons which leads to deficits in cognitive functions such as memory and thinking skills. Clinical diagnosis can be supported by biomarkers that detect the presence or absence of the disease. However, identifying such biomarkers, especially in a very early stage, remains challenging as brain changes due to AD occur even before amnestic symptoms appear [1]. The number of people diagnosed with dementia in the UK is expected to rise to over 2 million by 2051 with an estimated cost at between £17 billion and £18 billion a year (Dementia UK report[1]). Hence, identifying Alzheimer's disease (AD) earlier before the neurodegeneration is too severe and where treatment is not currently available, might aid in preventing AD onset. Specifically, patients initially diagnosed with mild cognitive impairment (MCI) are known to be a clinically heterogeneous group with different patterns of brain atrophy [2], of which some cases will not progress to AD [3]. To examine the borders between MCI and AD, Magnetic Resonance Imaging (MRI) was extensively used as a non-invasive imaging modality to track changes in brain images of MCI patients as they remain stable, progress to AD, or reverse to normal. Brain dementia MRI data are rapidly growing with emerging international research initiatives aiming to massively collect large high-quality brain images with structural, diffusion and functional imaging modalities, e.g., the public ADNI (Alzheimer's Disease Neuroimaging Initiative) dataset [4]. However, despite the large body of publications on AD and its early stages and major advances in neuroimaging technologies, brain image analysis and machine-learning methods, dementia research has not progressed as desired. Fundamentally, there are two major reasons for this.

First, the majority of methods developed for investigating AD stages have focused on learning how to *classify* AD vs. stable/progressive MCI or normal control (NC) subjects [5–26]. A conventional classification method would help identify features discriminating between MCI and AD groups; however, it would not allow to identify MCI patients with longer-term follow-up who will convert to AD after the first MR acquisition timepoint (i.e., baseline). Recently, a challenge on computer-aided diagnosis (CAD) of dementia based on structural MRI, namely CAD-Dementia [3], was launched to evaluate the performance of 29 algorithms from 15 research teams in classifying NC/MCI/AD using a public dataset. However, such dementia challenges have not focused on finding very early biomarkers of prodromal AD, characteristic of the presymptomatic MCI phase of the disease preceding severe cognitive decline, which is a major issue for current international research on AD.

Second, although advanced machine-learning and medical imaging analysis methods for dementia CAD have demonstrated high performance in the literature [3], they are not publicly shared for comparability, reproducibility, and gen-

---

[1] https://www.alzheimers.org.uk/about-us/policy-and-influencing/dementia-uk-report.

eralizability to unseen data [27]. Although the data for the CAD-dementia challenge is available, the developed methods were not made available for researchers to test on other datasets. A notable exception based on multivariate analysis [28] overlooks the richness and efficiency of recently published machine-learning and data analysis methods for brain disease diagnosis and prognosis [29]. In the following sections, we provide in-depth analysis of AD-related classification and evolution prediction methods from various neuroimaging modalities and identify the gaps in the state-of-the-art.

2   Selection Criteria

The included papers in this review were selected from MICCAI 2010-16 and IPMI 2011–17 proceedings. We conducted our search using different combinations of the following key words: *functional, structural, fMRI, DTI, magnetic resonance imaging, network, brain, connectivity, diffusion, Alzheimer's disease, mild cognitive impairment, classification, prediction, diagnosis, AD, MCI, biomarker, dementia.* We identified 37 papers based on the given search criteria (Fig. 1). It is noteworthy that works developed for segmentation tasks and those not focusing primarily on AD/MCI classification or MCI/AD evolution prediction were excluded from this review. We grouped them into two categories: (1) image-based methods and (2) network-based methods.

3   Image-Based Methods

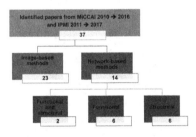

Fig. 1. Identified AD/MCI classification papers published in MIC-CAI 2010–2016 and IPMI 2011–2017 proceedings.

We identified 37 papers that use MR images for dementia state classification. These mainly used hippocampal atrophy and gray matter volume for classifying NC, MCI, and AD brain states. This can be explained by the fact that AD is related to gray matter loss [30] and the shape of subcortical structures (particularly the hippocampus) [31]. To predict clinical decline at the MCI stage and progression to AD, [23] created p-maps from the differences in the shape of the hippocampus between NC and AD subjects and showed increased rates by identifying local regions of interest (ROIs) within the hippocampus using statistical shape models. In a different work, [32] used (left and right) caudate nucleus, and putamen as additional features to hippocampal features to present a system for AD classification using a self-smoothing operator. Other papers suggested the combination of grading measure with hippocampal volume. Specifically, [5] proposed a new method to robustly detect hippocampal atrophy patterns based on a nonlocal means estimation framework. Combined with hippocampal volume, the grading measure (i.e, the atrophy degree in AD context) led to a success classification rate of 90% between NC and AD subjects.

[6] used two different modalities (MRI + PET) where each subject is represented by two 93-dimensional feature vectors that represent gray matter volume and the average intensity of PET images of 93 ROIs. A novel multi-task learning based feature selection method was proposed to preserve the complementary information conveyed by the two modalities and reached 94% accuracy in distinguishing between AD and NC subjects. In the same context, [8] proposed a manifold regularized multi-task learning framework to jointly select features from multi-modality (MRI + FDG-PET) data as well as [9] and [11,12,33]. Similarly, [7] used the same features in addition to three CSF biomarkers and introduced a deep learning method that discovers the non-linear correlations among features which improves the AD, MCI and MCI-C diagnosis accuracy. In [34], a novel Multifold Bayesian Kernelization (MBK) method was proposed to analyze multi-modal MRI biomarkers including average cerebral metabolic rate from PET data, gray matter volume, solidity and convexity features for AD and MCI classification. Another study [10] introduced a different approach to improve AD/NC and p-MCI/s-MCI classification. Basically, it learns a maximum margin representation using multiple atlases jointly with the classification model which resulted in 90% accuracy for AD/NC classification and 73% for p-MCI/s-MCI.

Another group of studies combined a different set of features with the image-based ones. For instance, [24] and [35] used cognitive scores including MMSE and ADAS in addition to volumes of brain structures and CSF features. [24] estimated disease progression in AD where a quantile regression was applied to learn two statistical models using (1) subjects that progress from a normal stage to MCI and (2) subjects that progress from MCI to AD. These models are then integrated into a multi-stage model for the whole disease course. On the other hand, to capture the associations between imaging and disease phenotype, [26] proposed a unified Bayesian framework based on genetic variants and image-based features. Additionally, [25] developed a supervised dimension reduction framework, called Spatially Weighted Principal Component Analysis (SWPCA), which integrates the spatial and graphic structure of imaging data into a statistical supervised learning paradigm (Fig. 2).

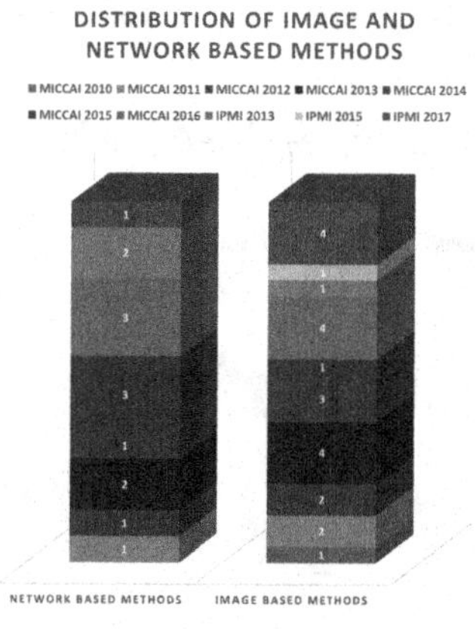

DISTRIBUTION OF IMAGE AND NETWORK BASED METHODS

■ MICCAI 2010 ■ MICCAI 2011 ■ MICCAI 2012 ■ MICCAI 2013 ■ MICCAI 2014
■ MICCAI 2015 ■ MICCAI 2016 ■ IPMI 2013 ■ IPMI 2015 ■ IPMI 2017

NETWORK BASED METHODS   IMAGE BASED METHODS

Fig. 2. Distributions of the identified image-based and network dementia diagnosis and prognosis methods.

We also identified two landmark papers [36,37], which devised machine learning frameworks for *predicting* dementia evolution at later timepoints. [37] proposed a novel canonical feature selection method to fuse information from different imaging modalities (MRI+PET). Specifically, original features (gray matter volume and average intensity of PET images) are projected into a common space. Hence, they become more comparable and easier to depict their relationship in order to predict clinical scores of Alzheimer's disease. Using the same features, [36] applied a low rank subspace clustering to cluster the data, then used a low rank matrix completion framework to identify pMCI patients and their time of conversion. In the same context, [38] extracted hippocampi from MRI scans and formulated an unsupervised framework for multi-task sparse feature learning based on a novel dictionary learning algorithm. This two-stage Multi-source Multi-Target Dictionary Learning (MMDL) algorithm demonstrated an improved prediction accuracy (with 2.61 rMSE in a timepoint of 12 months) in comparison with state-of-the-art methods.

4   Network-Based Methods

In this section, we identified 14 papers where 6 papers used functional networks (derived from fMRI) and another 6 used structural networks (derived from T1-weighted and diffusion MRI) for dementia diagnosis and prognosis. The remaining studies fused both networks to enhance AD classification accuracy.

Functional Networks. It has been reported that the neurodegenerative process of AD reflects disturbed functional connectivity between brain regions [39, 40]. These alterations are usually measured using resting-state functional MRI (rs-fMRI). Multiple methods have been used for AD diagnosis. For instance, [17] proposed a constrained sparse linear regression model associated with the least absolute shrinkage and selection operator (LASSO) that generates topologically consistent functional connectivity networks from rs-fMRI, thereby improving NC/MCI classification compared with traditional correlation-based methods [40, 41]. [18] introduced a sparse multivariate autoregressive (MAR) modeling to infer effective connectivity from rs-fMRI data and demonstrated its superiority compared to correlation based functional connectivity approaches.

[22] proposed a novel weighted sparse group representation method for brain network modeling, which integrates link strength, group structure as well as sparsity in a unified framework. This models the interactions among multiple brain regions unlike pairwise Pearson correlation and showed superior results in the task of MCI and NC classification. Since simply relying on pairwise functional connectivity between brain regions overlooks how their relationship might be affected at a higher-order level by AD, many studies introduced a functional connectivity hyper-network (FCHN) to infer additional information for AD classification. [19] constructed a FCHN using sparse representation modeling where three sets of brain regions are extracted to be fed into a multi-kernel SVM classifier and evaluated on a real MCI dataset. [42] also used a high-order functional connectivity network (HOFC) for MCI classification but generated multiple HOFC networks using hierarchical clustering to further ensemble them with a selective feature fusion method. This approach produced better classification performance than simple use of a single HOFC network.

Dynamic FC. While many studies assumed stationarity on the functional networks over time [43,44], recent studies in neuroscience have shown that functional organization changes spontaneously over time [45]. For instance, [20] introduced a novel method to model functional dynamics in rs-fMRI for MCI identification. Specifically, a deep network was designed to unravel the non-linear relationships among ROIs in a hierarchical manner and achieved a high classification performance.

Structural Networks. Recent findings showed that AD induces a disrupted topology in the structural network characterized by an early damage to synapses and a degeneration of axons [46]. These structural alterations can be pinned down using multimodal MRI imaging. In [13], a graph matching framework was devised to match (i) a source graph, where each node encodes a vector that describes regional gray matter volume or cortical thickness features, and (ii) a target graph that includes class label and clinical scores. This approach estimates a target vector for each sample without neglecting its relation with other samples. In similar context, [47] developed a matched signal detection (MSD) for graph-structure data under different signal models. This is particularly applied to the network classification of Alzheimer's disease where it achieved better performance than the traditional principal component analysis (PCA). [14] proposed a view-aligned hypergraph learning (VAHL) method using multi-modality data (MRI, PET, and CSF) for AD/MCI diagnosis where each view corresponds to a specific modality or a combination of several ones. This method can explicitly model the coherence among the views which led to a boost of 4.6% in classification accuracy. Another study [48] used also a novel dynamic hypergraph inference framework that works in a semi-supervised manner. This approach is flexible to integrate classification (identifying individuals with neuro-disease) and regression (predicting the clinical scores). [49] proposed a two-stage (query prediction + ranking) medical image retrieval technique with application to MCI diagnosis assistance. This framework was evaluated using three imaging modalities: T1-weighted imaging (T1-w), Diffusion Tensor Imaging (DTI) and Arterial Spin Labeling (ASL).

Functional and Structural Networks. Relying on either structural or functional brain networks may overlook the complementary information that can be leveraged to improve diagnosis and prognosis. For this purpose, [16] integrated two imaging modalities (DTI and fMRI) using a multi-kernel support vector machine (SVM) to improve classification performance where DTI images are parcellated into 90 regions. Then, different structural networks are generated, each conveying a different biophysical property of the brain (e.g., fibers count, fractional anisotropy and mean diffusivity). Additionally, functional connectivity matrices were constructed based on Pearson correlation coefficient to encode the connectivity strength between a pair of ROIs. Furthermore, [50] proposed a centralized hypergraph learning method to model the relationship among subjects using multiple MRIs. Specifically, four MRI sequences were used including T1-w MRI, DTI, rs-fMRI and ASL perfusion imaging. This allows to extract supplementary information captured by different neuroimaging modalities, thereby enhancing the quality of MCI diagnosis.

5   Results and Discussion

In this paper, we identified and reviewed 37 research papers on MCI and AD diagnosis and prognosis published in MICCAI 2010–2016 and IPMI 2011–2017 proceedings. Table 1 displays the different identified papers, while revealing five major issues that need to be addressed to move dementia research field forward.

First, all identified MICCAI papers focused on AD/MCI/NC classification, except for two papers [36,37], which proposed machine-learning frameworks to predict MCI conversion to AD. Similarly, the majority of the identified IPMI techniques tackled the problem of classification. We only were able to identify 4 IPMI papers [35,38,48,51] mostly focusing on predicting clinical cognitive scores (MMSA, ADAS) from a single or multiple timepoints. Undeniably, accurate discrimination between AD and MCI subjects is an important task to solve as it helps devise more individualized and patient-tailored treatment strategies [52]. However, an accurate prognosis for MCI patients is far more important for providing the optimal treatment and management of the disorder in very early stage. Indeed, early biomarkers identification might help reduce MCI to AD conversion rate. Therefore, predictive models need to be developed to fill this gap and propel the field of MCI prognosis forward. Such lack of studies could be due to the scarcity of spatiotemporal neuroimaging data where each patient is scanned multiple times. One way to tackle this is by adopting good practices in data analysis and sharing which can promote reliability and collaboration [27].

Second, the classification performance of the proposed technical methods for dementia largely varied with multiple peaks and drops as shown in Table 1. This can give insights into the heterogeneity and variability of the disease within subjects and how challenging it is to find an accurate method that works for all cases. In fact, no single approach can be sufficient as each has complementary merits and limitations.

Third, comparing these methods is very difficult since they used different approaches and datasets, it is somewhat hard to tell which one performs better if they are not compared against the same baseline methods and evaluated on the same dataset.

Fourth, all network-based analysis methods overlooked how dementia states affect the relationship between cortical regions *in morphology* in both stability, conversion, or reversal MCI evolution scenarios. To fill this gap and noting that several studies [29,53] reported that morphological features of the brain, such as cortical thickness, can be affected in neurological disorders, one can use the recently proposed morphological brain networks for dementia diagnosis [54,55].

Last, none of these works proposed a technique for predicting the developmental trajectory of brain shape changes as MCI progresses towards AD, remains stable, or reverses to normal. Besides, the absence of network-based predictive models is remarkable (Table 1). As such, the use of advanced network and shape

Table 1. *Identified brain dementia classification (diagnosis) and outcome prediction (prognosis) methods published in MICCAI 2010–2016 and IPMI 2011–2017 proceedings.* We reviewed pioneering seed works published at both leading conferences in medical image analysis, which typically get extended into journal papers. We report the classification/prediction accuracy for each paper. The majority of these works focused on classification tasks, e.g., classifying Alzheimer's disease (AD) patients vs. normal controls (NC) or stable/progressive mild cognitively impaired (sMCI/pMCI) vs. converted MCI (cMCI) patients. All these works proposed image-based or functional/structural network-based techniques. Notably, very few recent works from MICCAI [36,37] presented novel methods for prediction MCI late outcome using magnetic resonance images (MRI) acquired earlier before the outcome measurement. Another few IPMI studies [35,38,48,51] from IPMI focused mainly on predicting clinical cognitive score. More importantly, none of these works: (1) used morphological networks, which were recently introduced in [55] for early MCI diagnosis, or (2) proposed a technique for predicting the full trajectory of brain shape changes as MCI progresses towards AD, remains stable, or reverses to normal. These gaps need to be filled, considering there are studies that indicate morphological features of the brain, such as cortical thickness, can be affected in neurological disorders, including AD [29,53]. As such, the use of advanced network and shape analysis methods, using machine learning, could prove fruitful for both classification [54,55] and prediction tasks. AMP: Average Misclassification Percentage.

Data	Image-based papers	Network-based papers
AD/NC classification	[32,5,6,7,8,9,10,11,12,24,25,26] 96.25%—90%—94.37%—95.9%—95.03%—95.18%—90.69%—92.1%—96.1%—AMP(91%—86%)	[13,14,15,47] 92.17%—93.10%—94.05%—91%—100%
NC/MCI classification	[32,6,7,8,9,11,12,24] 91.25%—78.8%—85%—79.27%—79.52%—79.9%—80.3%—72%	[16,17,13,18,19,20,21,14,22,15] 96.59%—86.49%—81.57%—91.89%—94.6%—81.08%—84.85 %—80.00%—81.8 %—88.59%
cMCI/sMCI classification	[23,56,57,7,8,9,10,11,58,24] AUC(0.67)—69.4%—66%—75.8%—68.94%—72.02%—73.69%—80.7%—96.7%—78%	[14] 79%
AD/NC/MCI classification	[59,48] 48.48%—79.4% [34] NC: 86%—cMCI: 60.61%—sMCI: 66.96%—AD:81.76%	[51,48]
Prediction		
MCI conversion prediction using baseline MRI	[36,35] pMCI to AD: 76%—AUC(0.9)%	
MCI conversion prediction using MRI	[37] 18 months earlier: 76.55%; 12 months earlier: 79.83%	
MMSE and ADAS prediction	[38] rMSE (MMSE,ADAS-cog): 24 months earlier: (3.66,6.76); 18 months earlier: (3.37,4.64); 12 months earlier: (2.62,5.18)	[51,48] rMSE (MMSE,ADAS-cog): (1.74,4.12) — (1.62,3.47)

analysis methods, using machine learning, could prove fruitful for both classification [54,55] and prediction tasks.

6   Conclusion

Although not focusing on developing a cutting-edge methodology for advancing the field of medical data and information processing and analysis, we believe that this paper is of high merit as it is *the first* to review *seed cutting-edge neuroimaging-based methods published in MICCAI 2010–2016 and IPMI 2011–2017 proceedings* for Alzheimer's disease diagnosis and prognosis. We found that the majority of reviewed studies focused on NC, MCI and AD classification tasks using image-based methods or network-based methods including structural and functional brain networks. Although compelling, the majority of these works largely bypassed predictive models that could foresee the development of early AD stages including MCI over time. While the ultimate goal of classification is to provide a computer-aided diagnosis for better clinical decisions, predicting future progression of early demented brains from a baseline observation (i.e., a single timepoint) remains a priority as it might help delay conversion from MCI to AD when early treatment is addressed to the patient. Undoubtedly, predictive intelligence for early dementia diagnosis is still lagging behind, holding various untapped potentials for translational medicine.

References

1. Buckner, R.L.: Memory and executive function in aging and AD: multiple factors that cause decline and reserve factors that compensate. Neuron **44**, 195–208 (2004)
2. Misra, C., Fan, Y., Davatzikos, C.: Baseline and longitudinal patterns of brain atrophy in MCI patients, and their use in prediction of short-term conversion to AD: results from ADNI. NeuroImage **44**, 1415–1422 (2009)
3. Bron, E.E., et al.: Standardized evaluation of algorithms for computer-aided diagnosis of dementia based on structural MRI: the CADDementia challenge. NeuroImage **111**, 562–579 (2015)
4. Jack Jr., C.R., et al.: The Alzheimer's disease neuroimaging initiative (ADNI): MRI methods. J. Magn. Reson. Imaging: Official J. Int. Soc. Magn. Reson. Med. **27**, 685–691 (2008)
5. Coupé, P., Eskildsen, S.F., Manjón, J.V., Fonov, V., Collins, D.L.: Simultaneous segmentation and grading of hippocampus for patient classification with Alzheimer's disease. In: Fichtinger, G., Martel, A., Peters, T. (eds.) MICCAI 2011. LNCS, vol. 6893, pp. 149–157. Springer, Heidelberg (2011). https://doi.org/10.1007/978-3-642-23626-6_19
6. Liu, F., Wee, C.-Y., Chen, H., Shen, D.: Inter-modality relationship constrained multi-task feature selection for AD/MCI classification. In: Mori, K., Sakuma, I., Sato, Y., Barillot, C., Navab, N. (eds.) MICCAI 2013. LNCS, vol. 8149, pp. 308–315. Springer, Heidelberg (2013). https://doi.org/10.1007/978-3-642-40811-3_39
7. Suk, H.-I., Shen, D.: Deep learning-based feature representation for AD/MCI classification. In: Mori, K., Sakuma, I., Sato, Y., Barillot, C., Navab, N. (eds.) MICCAI 2013. LNCS, vol. 8150, pp. 583–590. Springer, Heidelberg (2013). https://doi.org/10.1007/978-3-642-40763-5_72

8. Jie, B., Zhang, D., Cheng, B., Shen, D.: Manifold regularized multi-task feature selection for multi-modality classification in Alzheimer's disease. In: Mori, K., Sakuma, I., Sato, Y., Barillot, C., Navab, N. (eds.) MICCAI 2013. LNCS, vol. 8149, pp. 275–283. Springer, Heidelberg (2013). https://doi.org/10.1007/978-3-642-40811-3_35

9. Suk, H.-I., Shen, D.: Clustering-induced multi-task learning for AD/MCI classification. In: Golland, P., Hata, N., Barillot, C., Hornegger, J., Howe, R. (eds.) MICCAI 2014. LNCS, vol. 8675, pp. 393–400. Springer, Cham (2014). https://doi.org/10.1007/978-3-319-10443-0_50

10. Min, R., Cheng, J., Price, T., Wu, G., Shen, D.: Maximum-margin based representation learning from multiple atlases for Alzheimer's disease classification. In: Golland, P., Hata, N., Barillot, C., Hornegger, J., Howe, R. (eds.) MICCAI 2014. LNCS, vol. 8674, pp. 212–219. Springer, Cham (2014). https://doi.org/10.1007/978-3-319-10470-6_27

11. An, L., Adeli, E., Liu, M., Zhang, J., Shen, D.: Semi-supervised hierarchical multimodal feature and sample selection for Alzheimer's disease diagnosis. In: Ourselin, S., Joskowicz, L., Sabuncu, M.R., Unal, G., Wells, W. (eds.) MICCAI 2016. LNCS, vol. 9901, pp. 79–87. Springer, Cham (2016). https://doi.org/10.1007/978-3-319-46723-8_10

12. Peng, J., An, L., Zhu, X., Jin, Y., Shen, D.: Structured sparse kernel learning for imaging genetics based Alzheimer's disease diagnosis. In: Ourselin, S., Joskowicz, L., Sabuncu, M.R., Unal, G., Wells, W. (eds.) MICCAI 2016. LNCS, vol. 9901, pp. 70–78. Springer, Cham (2016). https://doi.org/10.1007/978-3-319-46723-8_9

13. Liu, F., Suk, H.-I., Wee, C.-Y., Chen, H., Shen, D.: High-order graph matching based feature selection for Alzheimer's disease identification. In: Mori, K., Sakuma, I., Sato, Y., Barillot, C., Navab, N. (eds.) MICCAI 2013. LNCS, vol. 8150, pp. 311–318. Springer, Heidelberg (2013). https://doi.org/10.1007/978-3-642-40763-5_39

14. Liu, M., Zhang, J., Yap, P.-T., Shen, D.: Diagnosis of Alzheimer's disease using view-aligned hypergraph learning with incomplete multi-modality data. In: Ourselin, S., Joskowicz, L., Sabuncu, M.R., Unal, G., Wells, W. (eds.) MICCAI 2016. LNCS, vol. 9900, pp. 308–316. Springer, Cham (2016). https://doi.org/10.1007/978-3-319-46720-7_36

15. Liu, M., Du, J., Jie, B., Zhang, D.: Ordinal patterns for connectivity networks in brain disease diagnosis. In: Ourselin, S., Joskowicz, L., Sabuncu, M.R., Unal, G., Wells, W. (eds.) MICCAI 2016. LNCS, vol. 9900, pp. 1–9. Springer, Cham (2016). https://doi.org/10.1007/978-3-319-46720-7_1

16. Wee, C.-Y., Yap, P.-T., Zhang, D., Denny, K., Wang, L., Shen, D.: Identification of individuals with MCI via multimodality connectivity networks. In: Fichtinger, G., Martel, A., Peters, T. (eds.) MICCAI 2011. LNCS, vol. 6892, pp. 277–284. Springer, Heidelberg (2011). https://doi.org/10.1007/978-3-642-23629-7_34

17. Wee, C.-Y., Yap, P.-T., Zhang, D., Wang, L., Shen, D.: Constrained sparse functional connectivity networks for MCI classification. In: Ayache, N., Delingette, H., Golland, P., Mori, K. (eds.) MICCAI 2012. LNCS, vol. 7511, pp. 212–219. Springer, Heidelberg (2012). https://doi.org/10.1007/978-3-642-33418-4_27

18. Wee, C.-Y., Li, Y., Jie, B., Peng, Z.-W., Shen, D.: Identification of MCI using optimal sparse MAR modeled effective connectivity networks. In: Mori, K., Sakuma, I., Sato, Y., Barillot, C., Navab, N. (eds.) MICCAI 2013. LNCS, vol. 8150, pp. 319–327. Springer, Heidelberg (2013). https://doi.org/10.1007/978-3-642-40763-5_40

19. Jie, B., Shen, D., Zhang, D.: Brain connectivity hyper-network for MCI classification. In: Golland, P., Hata, N., Barillot, C., Hornegger, J., Howe, R. (eds.) MICCAI 2014. LNCS, vol. 8674, pp. 724–732. Springer, Cham (2014). https://doi.org/10.1007/978-3-319-10470-6_90

20. Suk, H.-I., Lee, S.-W., Shen, D.: A hybrid of deep network and hidden Markov model for MCI identification with resting-state fMRI. In: Navab, N., Hornegger, J., Wells, W.M., Frangi, A.F. (eds.) MICCAI 2015. LNCS, vol. 9349, pp. 573–580. Springer, Cham (2015). https://doi.org/10.1007/978-3-319-24553-9_70

21. Chen, X., et al.: High-order resting-state functional connectivity network for MCI classification. Hum. Brain Mapp. **37**, 3282–3296 (2016)

22. Yu, R., Zhang, H., An, L., Chen, X., Wei, Z., Shen, D.: Correlation-weighted sparse group representation for brain network construction in MCI classification. In: Ourselin, S., Joskowicz, L., Sabuncu, M.R., Unal, G., Wells, W. (eds.) MICCAI 2016. LNCS, vol. 9900, pp. 37–45. Springer, Cham (2016). https://doi.org/10.1007/978-3-319-46720-7_5

23. Leung, K.K., et al.: Increasing power to predict mild cognitive impairment conversion to Alzheimer's disease using hippocampal atrophy rate and statistical shape models. In: Jiang, T., Navab, N., Pluim, J.P.W., Viergever, M.A. (eds.) MICCAI 2010. LNCS, vol. 6362, pp. 125–132. Springer, Heidelberg (2010). https://doi.org/10.1007/978-3-642-15745-5_16

24. Schmidt-Richberg, A., et al.: Multi-stage biomarker models for progression estimation in Alzheimer's disease. In: Ourselin, S., Alexander, D.C., Westin, C.-F., Cardoso, M.J. (eds.) IPMI 2015. LNCS, vol. 9123, pp. 387–398. Springer, Cham (2015). https://doi.org/10.1007/978-3-319-19992-4_30

25. Guo, R., Ahn, M., Hongtu Zhu, H.Z.: Spatially weighted principal component analysis for imaging classification. J. Comput. Graph. Stat. **24**, 274–296 (2015)

26. Batmanghelich, N.K., Dalca, A.V., Sabuncu, M.R., Golland, P.: Joint modeling of imaging and genetics. In: Gee, J.C., Joshi, S., Pohl, K.M., Wells, W.M., Zöllei, L. (eds.) IPMI 2013. LNCS, vol. 7917, pp. 766–777. Springer, Heidelberg (2013). https://doi.org/10.1007/978-3-642-38868-2_64

27. Nichols, T.E., et al.: Best practices in data analysis and sharing in neuroimaging using MRI. Nat. Neurosci. **20**, 299 (2017)

28. Sabuncu, M.R., Konukoglu, E., Initiative, A.N., et al.: Clinical prediction from structural brain MRI scans: a large-scale empirical study. Neuroinformatics **13**, 31–46 (2015)

29. Brown, C., Hamarneh, G.: Machine learning on human connectome data from MRI. arXiv:1611.08699v1 (2016)

30. Karas, G., et al.: A comprehensive study of gray matter loss in patients with Alzheimer's disease using optimized voxel-based morphometry. Neuroimage **18**, 895–907 (2003)

31. Du, A., et al.: Magnetic resonance imaging of the entorhinal cortex and hippocampus in mild cognitive impairment and Alzheimer's disease. J. Neurol. Neurosurg. Psychiatry **71**, 441–447 (2001)

32. Iglesias, J.E., Jiang, J., Liu, C.-Y., Tu, Z.: Classification of Alzheimer's disease using a self-smoothing operator. In: Fichtinger, G., Martel, A., Peters, T. (eds.) MICCAI 2011. LNCS, vol. 6893, pp. 58–65. Springer, Heidelberg (2011). https://doi.org/10.1007/978-3-642-23626-6_8

33. Zhu, X., Suk, H.-I., Shen, D.: Multi-modality canonical feature selection for Alzheimer's disease diagnosis. In: Golland, P., Hata, N., Barillot, C., Hornegger, J., Howe, R. (eds.) MICCAI 2014. LNCS, vol. 8674, pp. 162–169. Springer, Cham (2014). https://doi.org/10.1007/978-3-319-10470-6_21

34. Liu, S., et al.: Multifold Bayesian kernelization in Alzheimer's diagnosis. In: Mori, K., Sakuma, I., Sato, Y., Barillot, C., Navab, N. (eds.) MICCAI 2013. LNCS, vol. 8150, pp. 303–310. Springer, Heidelberg (2013). https://doi.org/10.1007/978-3-642-40763-5_38

35. Venkatraghavan, V., Bron, E.E., Niessen, W.J., Klein, S.: A discriminative event based model for Alzheimer's disease progression modeling. In: Niethammer, M., et al. (eds.) IPMI 2017. LNCS, vol. 10265, pp. 121–133. Springer, Cham (2017). https://doi.org/10.1007/978-3-319-59050-9_10

36. Thung, K.-H., Yap, P.-T., Adeli-M, E., Shen, D.: Joint diagnosis and conversion time prediction of progressive mild cognitive impairment (pMCI) using low-rank subspace clustering and matrix completion. In: Navab, N., Hornegger, J., Wells, W.M., Frangi, A.F. (eds.) MICCAI 2015. LNCS, vol. 9351, pp. 527–534. Springer, Cham (2015). https://doi.org/10.1007/978-3-319-24574-4_63

37. Zhu, X., Suk, H.I., Lee, S.W., Shen, D.: Canonical feature selection for joint regression and multi-class identification in Alzheimer's disease diagnosis. Brain Imaging Behav. **10**, 818–828 (2016)

38. Zhang, J., Li, Q., Caselli, R.J., Thompson, P.M., Ye, J., Wang, Y.: Multi-source multi-target dictionary learning for prediction of cognitive decline. In: Niethammer, M., et al. (eds.) IPMI 2017. LNCS, vol. 10265, pp. 184–197. Springer, Cham (2017). https://doi.org/10.1007/978-3-319-59050-9_15

39. Fransson, P.: Spontaneous low-frequency bold signal fluctuations: an fMRI investigation of the resting-state default mode of brain function hypothesis. Hum. Brain Mapp. **26**, 15–29 (2005)

40. Wang, K., et al.: Altered functional connectivity in early Alzheimer's disease: a resting-state fMRI study. Hum. Brain Mapp. **28**, 967–978 (2007)

41. Wee, C.Y., et al.: Resting-state multi-spectrum functional connectivity networks for identification of MCI patients. PLoS ONE **7**, e37828 (2012)

42. Chen, X., Zhang, H., Shen, D.: Ensemble hierarchical high-order functional connectivity networks for MCI classification. In: Ourselin, S., Joskowicz, L., Sabuncu, M.R., Unal, G., Wells, W. (eds.) MICCAI 2016. LNCS, vol. 9901, pp. 18–25. Springer, Cham (2016). https://doi.org/10.1007/978-3-319-46723-8_3

43. Li, S., et al.: Analysis of group ICA-based connectivity measures from fMRI: application to Alzheimer's disease. PLoS ONE **7**, e49340 (2012)

44. Wee, C.Y., Yap, P.T., Zhang, D., Wang, L., Shen, D.: Group-constrained sparse fmri connectivity modeling for mild cognitive impairment identification. Brain Struct. Funct. **219**, 641–656 (2014)

45. Hutchison, R.M., et al.: Dynamic functional connectivity: promise, issues, and interpretations. Neuroimage **80**, 360–378 (2013)

46. Serrano-Pozo, A., Frosch, M.P., Masliah, E., Hyman, B.T.: Neuropathological alterations in Alzheimer's disease. Cold Spring Harb. Perspect. Med. **1**, a006189 (2011)

47. Hu, C., Cheng, L., Sepulcre, J., El Fakhri, G., Lu, Y.M., Li, Q.: Matched signal detection on graphs: theory and application to brain network classification. In: Gee, J.C., Joshi, S., Pohl, K.M., Wells, W.M., Zöllei, L. (eds.) IPMI 2013. LNCS, vol. 7917, pp. 1–12. Springer, Heidelberg (2013). https://doi.org/10.1007/978-3-642-38868-2_1

48. Zhu, Y., Zhu, X., Kim, M., Kaufer, D., Wu, G.: A novel dynamic hyper-graph inference framework for computer assisted diagnosis of neuro-diseases. In: Niethammer, M., et al. (eds.) IPMI 2017. LNCS, vol. 10265, pp. 158–169. Springer, Cham (2017). https://doi.org/10.1007/978-3-319-59050-9_13

49. Gao, Y., Adeli-M., E., Kim, M., Giannakopoulos, P., Haller, S., Shen, D.: Medical image retrieval using multi-graph learning for MCI diagnostic assistance. In: Navab, N., Hornegger, J., Wells, W.M., Frangi, A.F. (eds.) MICCAI 2015. LNCS, vol. 9350, pp. 86–93. Springer, Cham (2015). https://doi.org/10.1007/978-3-319-24571-3_11

50. Gao, Y., et al.: MCI identification by joint learning on multiple MRI data. In: Navab, N., Hornegger, J., Wells, W.M., Frangi, A.F. (eds.) MICCAI 2015. LNCS, vol. 9350, pp. 78–85. Springer, Cham (2015). https://doi.org/10.1007/978-3-319-24571-3_10

51. Wang, D., et al.: Structural brain network constrained neuroimaging marker identification for predicting cognitive functions. In: Gee, J.C., Joshi, S., Pohl, K.M., Wells, W.M., Zöllei, L. (eds.) IPMI 2013. LNCS, vol. 7917, pp. 536–547. Springer, Heidelberg (2013). https://doi.org/10.1007/978-3-642-38868-2_45

52. Ithapu, V.K., et al.: Imaging-based enrichment criteria using deep learning algorithms for efficient clinical trials in mild cognitive impairment. Alzheimer's Dementia **11**, 1489–1499 (2015)

53. Querbes, O., et al.: The Alzheimer's Disease Neuroimaging Initiative: early diagnosis of Alzheimer's disease using cortical thickness: impact of cognitive reserve. Brain **132**, 2036 (2009)

54. Mahjoub, I., Mahjoub, M.A., Rekik, I.: Brain multiplexes reveal morphological connectional biomarkers fingerprinting late brain dementia states. Sci. Rep. **8**, 4103 (2018)

55. Lisowska, A., Rekik, I.: Joint pairing and structured mapping of convolutional brain morphological multiplexes for early dementia diagnosis. Brain Connectivity **9**, 22–36 (2018)

56. Cheng, B., Zhang, D., Shen, D.: Domain transfer learning for MCI conversion prediction. In: Ayache, N., Delingette, H., Golland, P., Mori, K. (eds.) MICCAI 2012. LNCS, vol. 7510, pp. 82–90. Springer, Heidelberg (2012). https://doi.org/10.1007/978-3-642-33415-3_11

57. Singh, N., Wang, A.Y., Sankaranarayanan, P., Fletcher, P.T., Joshi, S.: Genetic, structural and functional imaging biomarkers for early detection of conversion from MCI to AD. In: Ayache, N., Delingette, H., Golland, P., Mori, K. (eds.) MICCAI 2012. LNCS, vol. 7510, pp. 132–140. Springer, Heidelberg (2012). https://doi.org/10.1007/978-3-642-33415-3_17

58. Zhang, J., et al.: Hyperbolic space sparse coding with its application on prediction of Alzheimer's disease in mild cognitive impairment. In: Ourselin, S., Joskowicz, L., Sabuncu, M.R., Unal, G., Wells, W. (eds.) MICCAI 2016. LNCS, vol. 9900, pp. 326–334. Springer, Cham (2016). https://doi.org/10.1007/978-3-319-46720-7_38

59. Wang, X., et al.: Predicting interrelated Alzheimer's disease outcomes via new self-learned structured low-rank model. In: Niethammer, M., et al. (eds.) IPMI 2017. LNCS, vol. 10265, pp. 198–209. Springer, Cham (2017). https://doi.org/10.1007/978-3-319-59050-9_16

Generative Adversarial Irregularity Detection in Mammography Images

Milad Ahmadi[1], Mohammad Sabokrou[2(✉)], Mahmood Fathy[1], Reza Berangi[1], and Ehsan Adeli[3]

[1] Iran University of Science and Technology (IUST), Tehran, Iran
milad_ahmadi@comp.iust.ac.ir
[2] Institute for Research in Fundamental Sciences (IPM), Tehran, Iran
sabokro@gmail.com
[3] Stanford University, Stanford, CA, USA

Abstract. This paper presents a novel method for detection of irregular tissues in mammography images. Previous works solved such irregularity detection tasks with binary classifiers. Here, we propose to detect irregularities by only observing the healthy samples and describe anything largely different from them as irregularity (i.e., unhealthy or cancerous tissues in terms of demographic breast images). This is particularly of great interest as it is very complicated to acquire datasets with all types of cancer cell shapes and tissues for building binary classifiers. Our modeling allows for learning an irregularity detector without any supervising signal from the irregular class. To this end, we propose an architecture with two deep convolutional networks (**R** and **M**) that are trained adversarially. **R** learns to Reconstruct regular mammography images by only observing healthy tissues and **M** (a Matching network) to detect if its input is healthy or not. The experimental results confirm the reliability and superior performance of our methods for detecting cancer tissues in mammography images in comparison with state-of-the-art irregularity detection methods. The code is available at https://github.com/milad-ahmadi/GAID.

1 Introduction

Invasive ductal carcinoma (also known as breast cancer) is the most common type of cancer threatening the lives of women worldwide [20]. Its early detection is very critical and one of the most important aspects of its treatment planning [8]. Mammography is the most common method for screening and diagnosing breast cancer before biopsy or utilizing any other types of imaging [25]. In standard breast examination procedures, X-Ray images form two different views are taken and one or two expert radiologists investigate these data. Exploring mammography images by experts is an expensive, challenging, and time-consuming task. Furthermore, due to the low contrast of such images, the interpretations may be troublesome from time to time.

© Springer Nature Switzerland AG 2019
I. Rekik et al. (Eds.): PRIME 2019, LNCS 11843, pp. 94–104, 2019.
https://doi.org/10.1007/978-3-030-32281-6_10

Previous work for detecting cancer in mammography images modeled the problem in a binary classification setting. Earlier methods developed approaches based on traditional machine learning techniques [6,14] and more recently deep learning methods have led to great improvements for detecting irregularities and delineating cancerous regions in mammography images [2,9,10,22]. However, cancer regions in mammography images are by nature irregularities materialized in between normal and healthy tissues. Modeling this as a two-class classification problem is ill-posed. Specifically for this problem, cancer regions may appear with different shapes or tissues and may have widely different levels of progression. Modeling this as a binary classification problem may lead to good results when used on datasets that contain only specific views and types of images with similar irregularity characteristics. We argue that the appropriate way is to learn what the healthy and normal tissues should look like and single out the abnormalities. On the other hand, manually labeling and annotating medical images is an expensive and time-consuming task. In this paper, we propose a method based on Generative Adversarial Networks (GAN) [5] with an end-to-end trainable architecture to identify irregularities in mammography images under unsupervised settings. Our proposed method is comprised of two networks, **R** and **M**, which are trained adversarially against each other. **R** is trained to reconstruct images while it fools **M** that its output does not contain any irregularity. In the meantime, **M** is trained to identify if its input is normal (healthy) tissue or is irregularity (i.e., cancer). After these two networks are trained, we define a scoring function, $S(\cdot)$, based on the outputs of both to quantify the likeliness of being irregularity for each input image. With this definition, our method is related to the recent advance in anomaly detection in computer vision applications, such as Adversarial Visual Irregularity Detection (AVID) [18], GANomaly [1], AnoGAN [19], and Adversarial Learned One-Class Classifier (ALOCC) [17]. Although these methods are designed for anomaly detection, they needed major modifications to accommodate the needs of our application. Specifically, ALOCC and GANomaly methods assumed that there exists a major difference between the concept of regular samples and irregular ones. This assumption is often reasonable for outlier detection and anomaly detection in RGB images. Medical images, however, are different and the difference between regular and irregular concepts is very minor. For instance, the noise element used in the previous works, such as ALOCC disrupts medical images. Also, as later shown by our experiments, GANomaly cannot properly reconstruct mammographic images and AnoGAN is too slow and has a hard time operating on limited training samples (since it exploits a residual loss function). In summary, the main contributions of this work are: (1) We propose an end-to-end trainable deep network able to detect irregularities (unhealthy cancerous tissues) in mammography images. To the best of our knowledge, this is the first approach for unsupervised irregularity detection in mammography images, i.e., the method detects irregular tissues in images without any supervising signal from the irregularity class. (2) Our method is tailored for detecting and localizing cancerous regions in mammography images. It obtains considerable performance which are comparable to fully-supervised methods and better than state-of-the-art irregularity detection methods, i.e., GANomaly [1], AnoGAN [19], and ALOCC [17].

2   Proposed Method

Detecting irregular tissues in mammography images can be defined as discovering regions that do not comply with normal (healthy) tissues present in the training set. To this end, we propose a method based on adversarial training composed of two important modules. The first module, denoted by $\mathbf{R}$, discovers the distribution of the healthy tissues by learning to reconstruct them. The second module, $\mathbf{M}$, learns to detect if its input is healthy or irregular (real or fake). In a nutshell, given an input image, $\mathbf{R}$ learns how to reconstruct it in a way that $\mathbf{M}$ does not identify it as irregular. Training $\mathbf{R}+\mathbf{M}$ in a joint adversarial manner, we propose a procedure to interpret mammography images and identify irregular cancerous regions. With respect to $\mathbf{R}$ and $\mathbf{M}$, we define an irregularity score function $\mathcal{S}(\cdot)$. The details of our Generative Adversarial Irregularity Detection (GAID) method are outlined in Fig. 1 and explained in the following subsections.

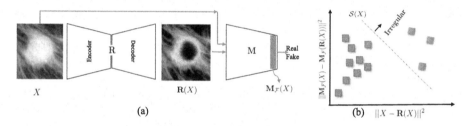

Fig. 1. Outline of proposed method: (a) $\mathbf{R}+\mathbf{M}$ are jointly learned to identify irregularity in mammography images. $\mathbf{R}$ learns to reconstruct healthy tissues. Whereas, $\mathbf{M}$ is trained to distinguish between regular (healthy) and irregular tissues (real and fake in GAN setting). During inference for a testing sample X, we use the final feature layer learned in $\mathbf{M}$, denoted by $\mathbf{M}_{\mathcal{F}}(X)$, to define the scoring function. (b) The scoring function $\mathcal{S}(X)$ is defined based on $||X - \mathbf{R}(X)||^2$ and $||\mathbf{M}_{\mathcal{F}}(X) - \mathbf{M}_{\mathcal{F}}(\mathbf{R}(X))||^2$.

R: Reconstruction Network: Encoder-decoder networks are widely used for tasks such as segmentation [3], inpainting [26], and anomaly detection [15–17]. Recently, similar structures were utilized for visual irregularity detection [18]. It was demonstrated that encoder-decoder networks can reconstruct normal samples present in the training set, when trained in an adversarial manner together with a discriminator. Similarly, we use an encoder-decoder network for $\mathbf{R}$, whose parameters, $\theta_{\mathbf{R}}$, are optimized to reconstruct healthy tissues. After training, $\mathbf{R}_{\theta_{\mathbf{R}}}(X)$ (or $\mathbf{R}(X)$ for short) should be similar or close to X. High deviations of $\mathbf{R}(X)$ with respect to X may conclude that X is not similar to the pool of training samples. Based on the previous work [1,18], $\mathbf{R}$ will be robust against some levels of noise in the input [23] while larger amounts of deviations will be considered as irregularity. We construct the training data such that it only involves images with healthy (normal) tissues. Therefore, **R** only learns to reconstruct healthy images, and will be incapable of reconstructing images with cancerous or irregular tissues. This incapability which is relative to irregularity likelihood can be measured by $||X - \mathbf{R}(X)||$. Figure 2(a) shows the details of **R**, which is composed of two sub-networks (1) encoder and (2) decoder. The encoder maps X to a latent space, which is then mapped back to the original image space using the decoder. For better stability, max-pooling layers are not used in this network, and instead kernel size of 5×5 with stride 2 are exploited in all convolutional layers. Except the last layer, Leaky ReLU activation function is applied on the output of convolutional layers of the encoder, and ReLU activation function on up-convolutional layers of the decoder. The last layer uses a $\tanh(\cdot)$ activation function.

M: Matching Network: As mentioned before, **R** is optimized to reconstruct regular tissues with a minimum error. Besides, **R** tries to fool **M** and trick it to accept its generated image as original and not fake. On the other hand, **M** focuses on discriminating between the reconstructed (fake) and original (real) images in the dataset. In some previous work [1,17], it is demonstrated that after the training of such discriminator in a GAN setting, it can detect irregular samples. As a result, similarly we train the the parameters of **M**, $\theta_\mathbf{M}$, which can distinguish between regular and irregular tissues at the inference time. In summary, if X contains irregular tissues, $\mathbf{M}_{\theta_\mathbf{M}}(X)$ (or $\mathbf{M}(X)$ for short) provides a good guidance to detect its irregularity. See Fig. 2(b) for the structure of **M**, which involves convolution and fully connected layers.

Irregularity Detector: Both **R** and **M** are trained to manipulate and analyze mammography images from different aspects. **R** learns to reconstruct its input and finally generate an output that not only has small reconstruction error, but it also can fool **M**. As a result, for input samples even with moderate to low amounts of noise, **R** can reconstruct them. But when there is an irregularity or high amounts of noise in the image (which were not present in the training images), **R** fails to reconstruct them and as a side effect decimates the input. Consequently, the reconstruction error is very high and hence it is a good discriminative measure for detecting irregular tissues in mammography images. On the other hand, **M** is learned to detect the irregular (fake) images. As shown by the previous work [17], after training the two networks in a similar setting as ours, $\mathbf{M}(X)$ defines a good detector for irregularities. However, we define a scoring function to reliably single out irregularities using both **R** and **M** trained networks.

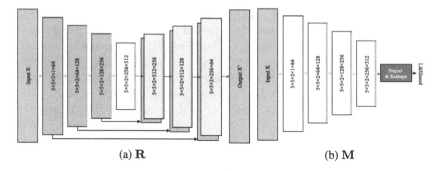

Fig. 2. Network structures. Batch normalization is included after some layers (denoted by the green bar). Every layer is defined using five elements (kernel-size-width × kernel-size-height × stride × input-feature-map-size × output-feature-map-size). (Color figure online)

To define this scoring function, suppose X and X' are normal and irregular samples, respectively. Based the on above intuitions, it is anticipated that $||X - \mathbf{R}(X)||^2 \leq ||X' - \mathbf{R}(X')||^2$. This means that the difference of an irregular sample compared to its reconstruction is more than the same difference metric for a regular sample. Let $\mathbf{M}_{\mathcal{F}}$ define the last fully connected layer of $\mathbf{M}$ (before the decision layer). Therefore, it is likely that $||\mathbf{M}_{\mathcal{F}}(X) - \mathbf{M}_{\mathcal{F}}(\mathbf{R}(X))||^2 \leq ||\mathbf{M}_{\mathcal{F}}(X') - \mathbf{M}_{\mathcal{F}}(\mathbf{R}(X'))||^2$. Based on these two useful directives from the $\mathbf{R}$ and $\mathbf{M}$ networks, a function for irregularity score can be defined as the following, such that a larger score will imply a larger likelihood for irregularity:

$$\mathcal{S}(X) = \lambda||X - \mathbf{R}(X)||^2 + (1 - \lambda)||\mathbf{M}_{\mathcal{F}}(X) - \mathbf{M}_{\mathcal{F}}(\mathbf{R}(X))||^2, \qquad (1)$$

where $\lambda \in [0, 1]$ is a regularization hyperparameter. With this scoring function, Eq. (2) summarizes the irregularity detector (ID) used for labeling of samples:

$$\text{ID}(X) = \begin{cases} \text{Regular tissue} & \text{if } \mathcal{S}(X) < \tau, \\ \text{Irregular tissue} & \text{otherwise,} \end{cases} \qquad (2)$$

where τ is a threshed value. subsectionTraining **R+M R+M** are adversarially trained on a training set containing only healthy mammography images. They are learned similar to GANs [5]. These two networks compete to each other, i.e., **R** tries to provide a lossless reconstruction from healthy images, and confuse **M** in distinguish between reconstructed and original images. Therefore, the training of **R+M** can be formulated as:

$$\min_{\mathbf{R}} \max_{\mathbf{M}} \left(\mathbb{E}_{X \sim p_{data}}[\log(\mathbf{M}(X))] + \mathbb{E}_{X \sim p_{data}}[\log(1 - \mathbf{M}(\mathbf{R}(X)))] \right), \qquad (3)$$

where p_{data} is the distribution of the training images (with healthy tissues). Parameters of $\theta_{\mathbf{R}}$ are optimized to efficiently reconstruct the images (with the loss term $\mathcal{L}_{rec}$) while being able to fool **M** that $\mathbf{R}(X)$ is a real image and not a reconstructed one ($\mathcal{L}_{real}$). Consequently, the loss function of **R** can defined as the combination of these two terms:

$$\mathcal{L}_{\mathbf{R}} = \underbrace{\beta_{rec}||\mathbf{R}(X) - X||}_{\mathcal{L}_{rec}(X)} + \underbrace{\beta_{real}f_\sigma\left(\mathbf{M}(\mathbf{R}(X)), 1\right)}_{\mathcal{L}_{real}(X)} \qquad (4)$$

where $f_\sigma(s, t)$ is sigmoid cross-entropy function, i.e., the cross-entropy over s with the desired label of t. β_{rec} and β_{real} are two hyperparameters to adjust the importance of each term (we set them equal to 1 and 4, respectively). On the other hand, **M** trains to identify $\mathbf{R}(X)$ as fake and X as real. Therefore, the objective function of **M**, $\mathcal{L}_{\mathbf{M}}$, is defined as:

$$\mathcal{L}_{\mathbf{M}} = \underbrace{f_\sigma(\mathbf{M}(\mathbf{R}(X)), 0)}_{\mathcal{L}_{fake}(X)} + \underbrace{f_\sigma(\mathbf{M}(X), 1)}_{\mathcal{L}_{real}(X)}. \qquad (5)$$

3 Experiments

In this section, we evaluate the proposed Generative Adversarial Irregularity Detection (GAID) method for detecting mammography image regions that contain irregular tissues. Comparison with the state-of-the-art in the scope of irregularity detection in mammography images is challenging due to two important difficulties: (1) Most of the previous methods solve this problem as a binary classification problem. This setting is different than ours while ours is more realistic, since obtaining enough irregular or cancerous samples may be very hard. Our method is the first that detects irregularities in mammography images by only observing and hence modeling healthy tissues; (2) The previous methods are often evaluated on private data, which make the comparison very complicated. To this end, for a fair set of comparisons, we re-implemented some binary classifiers on this dataset (AlexNet [11] and VGG-Net [24]) and compare with the fully-supervised settings, similar to the ones in [7]. Furthermore, we compare the performance of GAID with state-of-the-art anomaly detection methods (AnoGan [19], GANomaly [1], and ALOCC [17]) on the same mammography dataset. For a fair comparison among different methods, we execute them all with similar setting. We used the open-source shared source codes of these methods.

Experiment Setting: We implemented our method using of the Tensor-Flow framework on a 1080 Ti GPU. λ in the score function Eq. (1) is set to 0.4. All networks are learned with a batch size of 64. For the fully-supervised tests, since the number of irregular samples are very limited, available irregular patches are repeated to make equal sets of samples across the two classes. We Use the Contrast limited adaptive histogram equalization on mammography images for improving their contrast. We report the Area Under ROC Curve (AUC) and F1 score as evaluation metrics widely used by the previous works of irregularity detection. AUC is calculated by varying the τ threshold as introduced in (2). For our method and the other anomaly detection methods, the training set is composed of all normal samples (image patches), and there is no assumption about

Table 1. Comparison of results on the MIAS and INBreast datasets with the state-of-the-art methods for irregularity detection. In each column, the best results among of methods are typeset in boldface and the second best results are underlined. Complexity is in milliseconds of processing time per patch for processing of each of patch in the testing phase. [AUC/F1 Score]

Patch size	MIAS			INBreast			Run-time		
	64	128	256	64	128	256	64	128	256
AnoGAN [19]	0.66/0.61	0.60/0.55	0.64/0.57	0.78/0.73	0.80/0.76	0.68/0.60	4.13	10.31	37.42
GANomaly [1]	0.71/0.66	0.74/0.70	0.74/0.68	0.63/0.60	0.64/0.62	0.80/0.73	**0.31**	**0.67**	**3.18**
ALOCC [17]	0.70/0.65	0.74/0.67	0.72/0.66	0.70/0.67	0.78/0.74	0.75/0.71	3.45	9.02	32.40
GAID (Ours)	**0.72/0.67**	**0.76/0.71**	**0.76/0.70**	**0.74/0.69**	**0.87/0.80**	**0.86/0.79**	3.36	8.78	30.40

the context of irregular tissues during of training. For the binary classifiers, the irregular samples in the dataset are split equally between training and testing.

Datasets: *Mammographic Image Analysis Society (MIAS) Dataset:* This dataset [21] contains 322 mammography images in mediolateral-oblique (MLO) view with a 1024×1024 resolution. We consider all the benign and malignant cases in this dataset as irregular versus the normal class present in the dataset. The ground-truth of irregular (i.e., benign or malignant tumor) regions are annotated in the dataset using the center and the diameter of those regions. We partition all images into patches with three of size for different experiments (1) 217,866 patches of size 64×64, (2) 52,519 patches of size 128×128, and (3) 4,346 patches of size 256×256. These patches only contain healthy tissues and are used for training. For testing, in all three experiments 119 healthy and 119 irregular patches are considered (not included in the training set). Note that for building the supervised methods, 50% of irregularity patches are used for training.

INbreast Database: This dataset [13] contains 410 mammography images in mediolateral-oblique (MLO) and cranial-caudal (CC) views with a 3000×4000 resolution. We consider all the mass cases in this dataset as irregular versus the normal class present in the dataset.

We partition all images into patches with three of size for different experiments (1) 2,267,643 patches of size 64×64, (2) 541,341 patches of size 128×128, and (3) 135,155 patches of size 256×256. For testing, for different experiments (1) 177 healthy and 177 irregular (2) 1,596 healthy and 1,596 irregular (3) 10,259 healthy and 10,259 irregular.

Curated Breast Imaging Subset of Digital Database for Screening Mammography (CBIS-DDSM) Dataset: This dataset [12] contains 2,620 scanned film mammography studies from both cranial-caudal (CC) and mediolateral-oblique (MLO) views. The labels in this dataset also include benign, malignant, and normal with verified pathology information. We use this dataset only in a testing scenario and

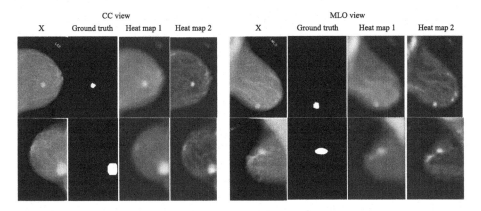

Fig. 3. Testing results of the proposed irregularity detector on the CBIS-DDSM dataset [12], trained on MIAS [21] and INBreast [13] datasets. Brighter areas of heat-map indicate higher likelihood of irregularity; The heat-map1 and heat-map2 are for training the GAID on MIAS and INBreast datasets, respectively.

Table 2. Comparison of results on the MIAS dataset with both fully-supervised baseline methods and the state-of-the-art methods for irregularity detection. 'Sup' stands for Supervision. [train with INBreast test with MIAS -best results]

	AlexNet	VGG-Net	AnoGAN [19]	GANomaly [1]	ALOCC [17]	GAID (Ours)
Sup	Full	Full	None	None	None	None
AUC	0.59	0.63	0.64	0.56	0.58	**0.70**
F1	0.54	0.60	0.62	0.50	0.53	**0.67**

qualitatively evaluate the pretrained model on MIAS on this data. This test can demonstrate the generalizability of the proposed method, specially when it is tested on an entirely new dataset.

Results: To evaluate the generalizability of GAID, we train it on MIAS [21] and INBreast, and test it on the CBIS-DDSM dataset [12]. Some of these results are visualized in Fig. 3. For each input image (X), two heat-maps are generated based on the irregularity score $\mathcal{S}(\cdot)$, defined in Eq. (1). Comparing the heat-maps with the ground-truth confirms that our method can delineate regions containing irregular tissues, i.e., cancerous regions. Note that the MIAS dataset contains images with only the CC view. This can be a reason why our method perform a bit worse on the images with the MLO view. Domain adaptation techniques can resolve such issues, which can define directions for future works.

Comparison with the State-of-the-Art Irregularity Detection Methods: Table 1 lists the comparison of our method with baseline and state-of-the-art methods on MIAS and INBreast datasets. In this table, the performance of the methods for detection of irregularities are provided in terms of AUC, F1-score, as well as their computational complexity (run-time at testing phase).

As can be seen, our method GAID outperforms all baseline and state-of-the-art methods, in some cases by a large margin. In terms of running time, our proposed method is better or as fast as other methods, except for GANomaly [1]. Considering both of aspects of run-time and detection accuracy, our method yields a good compromise. Note, patch size and number of training epochs are important for these results. Here, all models are trained with similar settings. Compared to the fully-supervised, although our method does not see any irregular samples during training, it obtains acceptable results. Note that accuracy scores of all unsupervised methods are significantly better than chance (i.e., p-value <0.01 in Fisher's exact test [4]). In this experiment, the supervised methods work well, as the diversity among the irregular samples in this dataset is rather low. On the other hand, if we test the trained models on entirely new datasets the supervised models perform worse in identifying the abnormalities. In real-world applications, imaging protocols and tumors can form in different shapes and with different progressing patterns. To show the generalizability of our method, we use the pretrained GAID and the supervised baseline, AlexNet, on the INBreast dataset, and test both on MIAS dataset [12]. Table 2 shows their performance.

4   Conclusion

In this paper, we proposed a simple yet effective method for detecting irregular tissues in mammography images. Our methods, GAID, was inspired by the adversarial learning techniques in the widely used GANs. We proposed two networks that were adversarially trained for interpreting mammography images. After training, based on these two networks, we defined an irregularity scoring function that was able to detect irregular tissues. Our method outperformed fully-supervised and the state-of-the-art methods for irregularity detection. Note, all previous methods for detecting irregular tissues in mammography were learned in a fully-supervised manner at the presence of both regular and irregular samples during training. Our method is the first that is trained without any supervision signal from the irregular class. In this work, our method operated on patches (i.e., it is a patch-based method). As a direction for the future work, one can extend the model to process the whole image at one-step, similar to AVID [18].

References

1. Akcay, S., Atapour-Abarghouei, A., Breckon, T.P.: GANomaly: semi-supervised anomaly detection via adversarial training. arXiv preprint arXiv:1805.06725 (2018)
2. Arevalo, J., González, F.A., Ramos-Pollán, R., Oliveira, J.L., Lopez, M.A.G.: Representation learning for mammography mass lesion classification with convolutional neural networks. Comput. Methods Programs Biomed. **127**, 248–257 (2016)
3. Fayyaz, M., Saffar, M.H., Sabokrou, M., Fathy, M., Huang, F., Klette, R.: STFCN: spatio-temporal fully convolutional neural network for semantic segmentation of street scenes. In: Chen, C.-S., Lu, J., Ma, K.-K. (eds.) ACCV 2016. LNCS, vol. 10116, pp. 493–509. Springer, Cham (2017). https://doi.org/10.1007/978-3-319-54407-6_33

4. Fisher, R.A.: The logic of inductive inference. J. Roy. Stat. Soc. **98**(1), 39–82 (1935)
5. Goodfellow, I., et al.: Generative adversarial nets. In: Advances in Neural Information Processing Systems, pp. 2672–2680 (2014)
6. Gubern-Mérida, A., et al.: Automated localization of breast cancer in DCE-MRI. Med. Image Anal. **20**(1), 265–274 (2015)
7. Huynh, B.Q., Li, H., Giger, M.L.: Digital mammographic tumor classification using transfer learning from deep convolutional neural networks. J. Med. Imaging **3**(3), 034501 (2016)
8. Jafari, S.H., et al.: Breast cancer diagnosis: Imaging techniques and biochemical markers. J. Cell. Physiol. **233**(7), 5200–5213 (2018)
9. Jamieson, A.R., Drukker, K., Giger, M.L.: Breast image feature learning with adaptive deconvolutional networks. In: Medical Imaging 2012: Computer-Aided Diagnosis, vol. 8315, p. 831506. International Society for Optics and Photonics (2012)
10. Jiao, Z., Gao, X., Wang, Y., Li, J.: A deep feature based framework for breast masses classification. Neurocomputing **197**, 221–231 (2016)
11. Krizhevsky, A., Sutskever, I., Hinton, G.E.: Imagenet classification with deep convolutional neural networks. In: Advances in Neural Information Processing Systems, pp. 1097–1105 (2012)
12. Lee, R.S., Gimenez, F., Hoogi, A., Miyake, K.K., Gorovoy, M., Rubin, D.L.: A curated mammography data set for use in computer-aided detection and diagnosis research. Sci. Data **4**, 170177 (2017)
13. Moreira, I.C., Amaral, I., Domingues, I., Cardoso, A., Cardoso, M.J., Cardoso, J.S.: INbreast: toward a full-field digital mammographic database. Acad. Radiol. **19**(2), 236–248 (2012)
14. Petrick, N., Chan, H.P., Sahiner, B., Wei, D.: An adaptive density-weighted contrast enhancement filter for mammographic breast mass detection. IEEE-TMI **15**(1), 59–67 (1996)
15. Sabokrou, M., Fathy, M., Hoseini, M.: Video anomaly detection and localisation based on the sparsity and reconstruction error of auto-encoder. Electron. Lett. **52**(13), 1122–1124 (2016)
16. Sabokrou, M., Fayyaz, M., Fathy, M., Klette, R.: Deep-cascade: cascading 3D deep neural networks for fast anomaly detection and localization in crowded scenes. IEEE Trans. Image Process. **26**(4), 1992–2004 (2017)
17. Sabokrou, M., Khalooei, M., Fathy, M., Adeli, E.: Adversarially learned one-class classifier for novelty detection. In: Proceedings of the IEEE Conference on Computer Vision and Pattern Recognition, pp. 3379–3388 (2018)
18. Sabokrou, M., et al.: AVID: adversarial visual irregularity detection. arXiv preprint arXiv:1805.09521 (2018)
19. Schlegl, T., Seeböck, P., Waldstein, S.M., Schmidt-Erfurth, U., Langs, G.: Unsupervised anomaly detection with generative adversarial networks to guide marker discovery. In: Niethammer, M., et al. (eds.) IPMI 2017. LNCS, vol. 10265, pp. 146–157. Springer, Cham (2017). https://doi.org/10.1007/978-3-319-59050-9_12
20. Stewart, B., Wild, C.P., et al.: World cancer report 2014. Self (2018)
21. Suckling, J., et al.: Mammographic image analysis society (MIAS) database v1.21 [dataset] (2015). https://www.repository.cam.ac.uk/handle/1810/250394
22. Sun, W., Tseng, T.L.B., Zhang, J., Qian, W.: Enhancing deep convolutional neural network scheme for breast cancer diagnosis with unlabeled data. Comput. Med. Imaging Graph. **57**, 4–9 (2017)

23. Vincent, P., Larochelle, H., Bengio, Y., Manzagol, P.A.: Extracting and composing robust features with denoising autoencoders. In: Proceedings of the 25th International Conference on Machine Learning, pp. 1096–1103. ACM (2008)
24. Wang, L., Guo, S., Huang, W., Qiao, Y.: Places205-VGGNet models for scene recognition. arXiv preprint arXiv:1508.01667 (2015)
25. Weissleder, R.: Molecular imaging in cancer. Science **312**(5777), 1168–1171 (2006)
26. Xie, J., Xu, L., Chen, E.: Image denoising and inpainting with deep neural networks. In: Advances in Neural Information Processing Systems, pp. 341–349 (2012)

Hierarchical Adversarial Connectomic Domain Alignment for Target Brain Graph Prediction and Classification from a Source Graph

Alaa Bessadok[1,2], Mohamed Ali Mahjoub[1], and Islem Rekik[2(✉)]

[1] LATIS Lab, ISITCOM, University of Sousse, Sousse, Tunisia
[2] BASIRA Lab, Faculty of Computer and Informatics,
Istanbul Technical University, Istanbul, Turkey
irekik@itu.edu.tr
http://basira-lab.com/

Abstract. Recently, deep learning methods have been widely used for medical data synthesis. However, existing deep learning frameworks are mainly designed to predict Euclidian structured data (i.e., image), which causes them to fail when handling *geometric* data (e.g., brain graphs). Besides, these do not naturally account for domain fracture between training source and target data distributions and generally ignore any hierarchical structure that might be present in the data, which causes a *flat domain alignment*. To address these limitations, we unprecedentedly propose a Hierarchical Graph Adversarial Domain Alignment (HADA) framework for predicting a target brain graph from a source brain graph. We first propose to align the source domain to the target domain by learning their successive embeddings using training samples. In this way, we are optimally learning a *hierarchical alignment* of both domains as we are learning a graph embedding using the previously aligned source-to-target embedding. To predict the target brain graph of a testing subject, we learn a source embedding using training and testing samples. Second, we learn a connectomic manifold for each of the resulting embeddings (i.e., the hierarchical embedding and the source embedding). Next, we select the nearest neighbors to the testing subject in the source manifold in order to average their corresponding target graphs. Finally, using the source and *predicted* target graphs by our HADA method, we train a random forest classifier to distinguish between disordered and healthy subjects. Our HADA framework outperformed comparison methods in predicting target brain graph and yields the best classification accuracy in comparison with using only the source graphs.

This work was supported by BASIRA Talented Minority Scholarship for research students in low research & development countries http://basira-lab.com/prime-miccai19/.

I. Rekik et al. (Eds.): PRIME 2019, LNCS 11843, pp. 105–114, 2019.
https://doi.org/10.1007/978-3-030-32281-6_11

1   Introduction

In the last decade, synthesizing medical images has spanned different works to circumvent the high costs of acquiring medical scans such as magnetic resonance imaging (MRI). Most of the existing works aiming to predict a medical image from another are deep learning based approaches. For instance, [1] proposed to predict positron-emission tomography (PET) image from an MRI image using convolutional neural networks (CNN). Later on, [2] leveraged a conditional Generative Adversarial Network (GAN) [3] along with a fully convolutional network to predict PET from computerized tomography (CT) scans. However, these methods focus only on synthesizing Euclidian structured data (i.e., image modalities). Hence, they might fail in handling non-Euclidian *geometric data* such as graphs and manifolds. Recently, deep learning methods were trained on brain graphs to diagnose neurological diseases. For example, based on functional brain graphs extracted from resting state fMRI (rs-fMRI) data, [4] and [5] used Graph Convolution Network (GCN) [6] for gender classification and Autism Spectrum Disorder (ASD) diagnosis, respectively. However, to date, most of these geometric deep learning (GDL) frameworks overlook the problem of *graph synthesis*. Particularly, predicting a target brain graph from a source graph remains largely unexplored.

However, making such an inter-domain prediction is challenging due to the domain fracture problem resulting in the difference in distribution between the source and target domains. Existing works proposed GAN-based frameworks to overcome the fracture between two domains. For example, [7] used a typical GAN to align T1- and T2-weighted MRI images to Magnetic Resonance Angiography (MRA) scans. In [8], a cycle-consistent GAN was used for a bi-directional domain alignment where they first mapped the CT source domain to the MRI target domain and then learned the reverse mapping. Specifically, assuming that cycleGAN does not have a constraint between the generated target image and its ground truth target image, they added a structure-consistency loss to the original cycleGAN and adopted it to predict MRI data from CT data. However, one major limitation of these GAN-based methods lies in synthesizing medical images rather than geometric medical data such as brain graph. To the best of our knowledge, no previous work used geometric deep learning to predict brain graph from a source graph [9]. A second major limitation is that they learn a *flat domain alignment* as they only move the source domain to the target domain and are unable to learn the hierarchical structure that exists within the resulting aligned domains.

In order to solve the above challenges, we propose a Hierarchical Adversarial Domain Alignment (HADA) architecture, which predicts a target brain graph from a single source graph while hierarchically aligning both domains. Specifically, we leverage the adversarially-regularized generative autoencoder (ARGA) proposed in [10] which is a GAN-based method extended to graphs. ARGA comprises a generator G defined as a GCN [6] and a discriminator D_{align} defined as a multilinear perceptron. However, ARGA is mainly devised for graph embedding task and not for graph prediction. Additionally, it makes an *intra-domain*

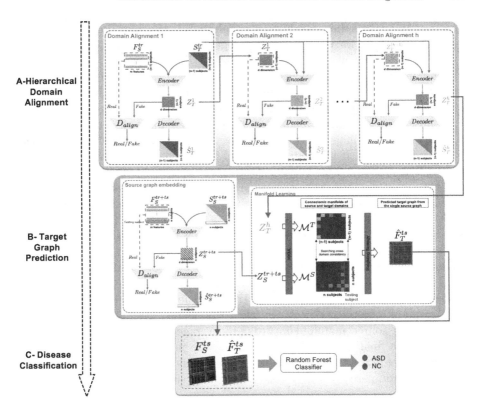

Fig. 1. *Proposed framework of Hierarchical Adversarial Domain Alignment (HADA) for target brain graph prediction from a single source graph to boost disease diagnosis.* **A- Hierarchical Domain Alignment.** We train h source to target domain alignment models in a hierarchical manner. At the baseline level, we align the source brain graphs to the target brain graphs. At the current level h, we learn the embedding of the aligned source-to-target graphs using the embedding produced by the previous model at level $(h-1)$. All domain alignment models are regularized by a discriminator D_{align} to align the distribution of the learned embeddings with the prior distribution of the training samples. **B- Target Graph Prediction.** We first learn a source embedding of training and testing brain graphs. Then, we learn a target manifold that nests the hierarchically aligned and embedded graphs of the training subjects, and a source manifold that nests the embedded source graphs of the training and testing subjects. To predict the final testing target graph, we select the training subjects (i) most similar to the testing subject in the aligned source manifold and (ii) with highest cross-domain score quantifying the overlap in their local neighborhoods across source and target domains. Next, we average their corresponding target graphs to predict the missing target brain graph. **C- Disease Classification.** We train a random forest classifier using concatenated ground-truth source and predicted target graphs to distinguish between autism spectrum disorder (ASD) and normal controls (NC).

alignment as it learns the embedding of graphs within the same domain. Here, we propose to extend it for jointly solving the graph prediction and the hierarchical *inter-domain* alignment.

Fundamentally, our HADA framework has three main contributions: (i) the hierarchical adversarial domain alignment, (ii) the target brain graph prediction and (iii) classification while leveraging a single source graph. In the first step, we propose an inter-domain alignment using multiple and consecutive levels. At a baseline level, we map the original source brain graphs to the target graphs. At the current level h we map the embedded graphs learned at the previous level $(h-1)$ to the target graphs. Thus, each layer of HADA gradually captures the local structure of the aligned graphs. Hence, it generates a hierarchical low-dimensional representation (i.e., embedding) of the aligned source to target graphs. This alignment is regularized using a discriminator D_{align}. In the second step, we first learn a source embedding using training and testing graphs. Then, we propose to learn two connectomic manifolds: the first one captures the similarities between the embedded source graphs of both training and testing subjects, and the second one captures the similarities between the hierarchically aligned source-to-target graphs of the training subjects. To predict the target brain graph of a testing subject, we select the nearest neighbors to the testing sample in the source manifold in order to average their corresponding target graphs. In the final step, we propose to use the predicted target graph along with the source graph to further improve the classification accuracy of a random forest model compared where trained used only a single source graph.

2   Proposed HADA for Graph Prediction and Classification

In this section, we detail our joint graph prediction and hierarchical domain alignment framework. We illustrate in Fig. 1 our three proposed steps: (1) hierarchical adversarial domain alignment of source and target brain graphs, (2) prediction of target brain graph, and (3) disease classification using the predicted target graphs.

2.1   Hierarchical Adversarial Domain Alignment of Source and Target Domains

Each subject in our dataset has a source brain graph and a target graph. Each graph is encoded in a symmetric matrix as it captures the morphological relationship between anatomical regions extracted from the parcellation of the cortical surface [11–14]. We extract the upper-diagonal part of each matrix to create a graph feature vector. Then, we stack the feature vectors of all disordered and healthy subjects vertically to get $\mathbf{F}_S$ matrix of the source graphs and $\mathbf{F}_T$ matrix of target graphs, each has a size of $(n \times m)$ and $((n-1) \times m)$, respectively. We denote n the total number of subjects in the dataset and m the number of features. We aim in this step to align the source feature vectors to the target vectors while learning the hierarchical embedding of the aligned source-to-target graphs. Specifically, we propose to learn multiple and successive graph embeddings where each embedding depends to the previous one. To achieve this, we use a GAN-based method dedicated to graph embedding called ARGA, that is a graph convolutional autoencoder regularized by a discriminator. It is composed of a GCN encoder (i.e., generator) that learns the latent representation of the brain graphs while taking as input the feature matrix and the adjacency matrix capturing the topological structure of the graph. Basically, we stack h ARGA models in a hierarchical way to align the source to the target training graphs (Fig. 1–A). The first model takes as input the original source graphs $\mathbf{F}_S^{tr}$ of the training subjects in addition to the adjacency matrix $\mathbf{S}_T^{tr}$ that encodes the similarities between training subjects using their target graphs. It is also given as input to all h generators. Besides, we propose to learn the graph adjacency matrix using MKML algorithm [15]. Instead of using ordinary distance metric (e.g., Euclidean distance) that might fail to capture heterogeneous data distribution this algorithm adopted multiple kernels (i.e., distance metric) to learn the similarity between data points with high dimensionality and heterogeneous distribution. The resulting aligned source-to-target embedding $\mathbf{Z}_{S \to T}^{(h-1)}$ of this level is given as input to the following alignment level h. Thus, we define our generators at two consecutive levels $(h - 1)$ and h as follows:

$$\mathbf{Z}_{S \to T}^{(h-1)} = G^{(h-1)}(\mathbf{F}_S^{tr}, \mathbf{S}_T^{tr}); h = 2;$$
$$\mathbf{Z}_{S \to T}^{(h)} = G^{(h)}(\mathbf{Z}_{S \to T}^{(h-1)}, \mathbf{S}_T^{tr}); h \succeq 2$$

$\mathbf{Z}_{S \to T}^{(h-1)}$ is the embedding of the aligned source-to-target domains of the previous level $(h - 1)$. Furthermore, each GCN encoder used in our framework is constructed with two layers. We denote the input feature matrix at the first layer as $\mathbf{X}$. At a baseline level, we initialize $\mathbf{X} = \mathbf{F}_S^{tr}$, and at the following consecutive levels we define it as $\mathbf{X} = \mathbf{Z}_{S \to T}^{(h-1)}$. Additionally, we denote the input adjacency matrix for both GCN layers as $\mathbf{S}$ and denote the filter to learn the convolution of the layers l as $\mathbf{W}^{(l)}$. Given these inputs, we apply the two following equations:

$$\mathbf{Z}^{(1)} = f_{ReLU}(\mathbf{X}, \mathbf{S}|\mathbf{W}^{(0)}); \mathbf{Z}^{(2)} = f_{linear}(\mathbf{Z}^{(1)}, \mathbf{S}|\mathbf{W}^{(1)}),$$

$\mathbf{Z}^{(1)}$ and $\mathbf{Z}^{(2)}$ are the encoder results of the first and the second GCN layers, respectively. $\mathbf{Z}^{(2)}$ represents the desired aligned source-to-target embeddings at a specific alignment level h. Rectified Linear Unit, $ReLU(.)$ is the activation function of the first layer and a linear function is used for the second layer. The function $f_{(.)}$ is defined as follows:

$$f_\phi(\mathbf{X}^{(l)}, \mathbf{S}|\mathbf{W}^{(l)}) = \phi(\widetilde{\mathbf{D}}^{-\frac{1}{2}}\widetilde{\mathbf{S}}\widetilde{\mathbf{D}}^{-\frac{1}{2}}\mathbf{X}^{(l)}\mathbf{W}^{(l)})$$

ϕ is the activation function for a specific layer (l), $\widetilde{\mathbf{S}} = \mathbf{S} + \mathbf{I}$ where $\mathbf{I}$ is the identity matrix, and $\widetilde{\mathbf{D}}_{ii} = \sum_j \widetilde{\mathbf{S}}_{ij}$ is a diagonal matrix. We propose to decode the embedded graph $\mathbf{Z}$ by reconstructing the target adjacency matrix $\hat{\mathbf{S}}$. Specifically, we compute the sigmoid function of the dot product of embedded graphs $\mathbf{z}_j$ of the subject (i.e., node) j and the transposed embedded graphs $\mathbf{z}_i^\top$ of the subject i.

$$Dec(\hat{\mathbf{S}}|\mathbf{Z}) = \frac{1}{1 + e^{-(\mathbf{z}_i^\top \cdot \mathbf{z}_j)}}$$

At the baseline level, HADA acts as a typical ARGA where the distribution of the learned source embeddings is aligned with the prior distribution of the input source graphs. In the next levels, HADA hierarchically moves the source graph distribution towards the target graph distribution via a nested embedding strategy where the learning of the current source-to-target embedding $\mathbf{Z}_{S \to T}^{(h)}$ at level h depends on the previously learned embedding $\mathbf{Z}_{S \to T}^{(h-1)}$ at level $(h - 1)$. This level-wise source-to-target alignment is reinforced by an adversarial regularizer D_{align} that shifts the learned embedding distribution towards the prior data distribution. This discriminator is a multi-layer perceptron, considered as a binary classifier, that distinguish between real data distribution (source graphs at the first level or the embeddings of the aligned source-to-target graphs at the ongoing levels) and fake data distribution generated from our encoder G. Hence, we formulate the cost function at level h of the adversarial alignment of source and target domains as follows:

$$\min_{G^{(h)}} \max_{D_{align}} \mathbf{E}_{p(real)}[\log D_{align}(\mathbf{Z}^{(h)})] + \mathbf{E}_{p(fake)}[\log(1 - D_{align}(G^{(h)}))]$$

where $\mathbf{E}$ is the cross-entropy cost, $G^{(h)}$ is the graph encoder of the alignment level h and D_{align} is the binary classifier with maximum log likelihood objective.

2.2   Target Graph Prediction for the Testing Subject

To predict the target graph of a testing subject, we propose to select its nearest training neighbors in the source domain. Next, we average their corresponding *hierarchically aligned* target graphs, thereby predicting the missing target graph. To achieve this, we learn another generator $G_S(\mathbf{F}_S^{tr+ts}, \mathbf{S}_S^{tr+ts})$ for source graph embedding of training and testing subjects (Fig. 1–B). Specifically, $\mathbf{F}_S^{tr+ts}$ denotes the stacked source feature vector of the training and testing subjects, and $\mathbf{S}_S^{tr+ts}$ denotes the learned similarity matrix [15] using their source graphs. As the graphs used in this step rely to a single domain (i.e., source domain), we perform a flat domain alignment by learning a source graph embedding. We propose to regularize the resulting latent representation of source domain using the discriminator D_{align}. In particular, we define the energy function of this source embedding model as follows:

$$\min_{G_S} \max_{D_{align}} \mathbf{E}_{p(real)}[\log D_{align}(\mathbf{Z}_S^{tr+ts})] + \mathbf{E}_{p(fake)}[\log(1 - D_{align}(G_S(\mathbf{F}_S^{tr+ts}, \mathbf{S}_S^{tr+ts})))]$$

Next, we use the learned source embeddings $\mathbf{Z}_S^{tr+ts}$ to find the most similar training subjects to the testing subject. Mainly, we propose to enforce a local consistency in source and target neighborhoods for the selected training source neighbors. To further boost the reliability of the selected training source sample we select those having a large overlap in nearest neighbors across the embedded source $\mathbf{Z}_S^{tr+ts}$ and aligned source-to-target domains $\mathbf{Z}_{S \rightarrow T}^{(h)}$. Using MKML [15], we learn a connectomic manifold $\mathcal{M}^S$ that quantifies the relationship between training and testing subjects using their source embeddings, and learn a second connectomic manifold $\mathcal{M}^T$ using the hierarchically aligned source-to-target embeddings of the training subjects (Fig. 1–B). Then, we identify in $\mathcal{M}^S$ the top K-closest training subjects to the testing subject and find for each of these K selected training samples, its nearest neighbors in both manifolds $\mathcal{M}^S$ and $\mathcal{M}^T$. Next, we extract a list $\mathbf{L}_S$ of its top m closest neighbors in $\mathcal{M}^S$ and $\mathbf{L}_T$ in $\mathcal{M}^T$ in order to assign a weighted similarity score $w(k)$ for each training subject k. Finally, we compute their overlap using the following formula: $\mathbf{w}(k) = \exp\left(\left(\frac{\mathbf{L}_S \cap \mathbf{L}_T}{m}\right) \times \mathcal{M}^S(ts, k)\right)$. In order to estimate the target graph of the testing subject, we average the target graphs of the selected m neighbors with the highest w scores.

2.3   Brain Graph Classification

To evaluate the reliability of our predicted target brain graphs for diagnosing neurological disorders using minimal connectomic data (i.e., single brain network), we propose to use a random forest (RF) [16] binary classifier to classify subjects as disordered or healthy. Instead of training the RF model using only the available source graphs, we propose to further integrate the predicted target graphs in the classifier training step. For each subject, we concatenate horizontally the source and the predicted target feature vectors. Then, we use our RF model to classify the disordered and healthy subjects using leave-one-out cross-validation (Fig. 1–C).

3   Results and Discussion

Dataset and Model Configuration. We used leave-one-out cross-validation to evaluate our proposed framework on 75 ASD and 75 NC subjects from the Autism Brain Imaging Data Exchange (ABIDE[1]) public dataset, each with structural T1-w MRI data. Each subject has two morphological brain graphs constructed using the following cortical measurements: maximum principal curvature and the average curvature. Both hemispheres were reconstructed using FreeSurfer [17], and parcellated into 35 anatomical regions of interest (ROI) using Desikan-Killiany Atlas. The extracted ROIs represent the nodes in the brain graphs, and the edges encode the morphological dissimilarity between pairs of ROIs. We construct our encoder with two layers, the hidden and the embedding layers, each comprising 32 neurons. We construct our discriminator D_{align} with two layers composed of 64 and 32 neurons, respectively. We fix both learning rates of the encoder and discriminator to 0.001 as in [10]. For MKML parameters [15], we set the number of kernels to 10 and the number of cluster to 3.

---

[1] http://fcon_1000.projects.nitrc.org/indi/abide/.

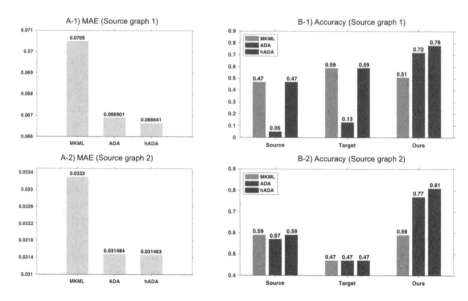

Fig. 2. *Comparison of HADA against MKML and ADA methods using mean absolute error (MAE) and classification accuracy.* Source graph 1: morphological brain network derived from the maximum principal curvature. Source graph 2: morphological brain network derived from average curvature. Source: classification of samples using their ground-truth source brain graphs. Target: classification of samples using their original target graphs. Ours: the concatenated ground-truth source and predicted target feature vectors used for the classification task. MKML: the target graph prediction without domain alignment using manifold learning [15]. ADA: the flat adversarial domain alignment using ARGA [10]. HADA: the proposed hierarchical adversarial regularization for source-to-target domain alignment.

For target graph prediction, we set the number of source neighbors to 5. For the random forest classifier, we use a grid search to identify the optimal number of decision trees $(200, \ldots, 2000)$, the minimum number of samples to split a node $(2, 5, 10)$ and the number of samples at a leaf node $(1, 2, 4)$.

Comparison Methods and Evaluation. We evaluate the benefit of HADA against two baseline methods: **(1) MKML:** it is a variant of the target graph prediction step illustrated in (Fig. 1–B) where the connectomic manifolds [15] are learned using the original brain graphs. Specifically, we predict the target graph without performing any domain alignment step. **(2) ADA:** we use for this method the original ARGA [10] for a flat adversarial domain alignment, where we use one level to move the source domain to the target one. Our experiments explore two different applications. In a first evaluation, we use mean absolute error (MAE) to measure the prediction error between the ground truth graphs and their corresponding predicted graphs. Figure 2–A-1 and A-2 show the lowest MAE produced by HADA when considering maximum principal curvature and the average curvature as the source graphs, respectively. This demonstrates that our proposed hierarchical domain alignment boost brain graph prediction accuracy from a source graph. In a second evaluation, we report the classification accuracy of each comparison method in addition to our HADA using the source and target graphs independently, and the concatenated source and predicted target graphs. Figure 2–B-1 and B-2 illustrate that our HADA achieved the highest classification accuracy when including the predicted target graphs with the source graphs. This clearly demonstrates that predicted target graphs produced by our hierarchical domain alignment framework boost the disease classification task.

4   Conclusion

We proposed a novel geometric deep learning framework for predicting a brain graph from a single source graph. Our key contribution consists in designing a *hierarchical domain alignment* of source graphs to the target graphs by successively learning their latent representations. Thus, our model can capture the hierarchical structure existing in both source and target graphs. Experiments on the autism dataset showed that our proposed HADA is able to achieve a low prediction error and a high classification accuracy compared to benchmark methods. Interesting future directions include leveraging our proposed hierarchical adversarial domain alignment for jointly predicting multimodal target graphs (e.g., functional and structural brain graphs) from a single source graph and comparing our model to existing domain adaptation methods.

References

1. Li, R., et al.: Deep learning based imaging data completion for improved brain disease diagnosis. In: Golland, P., Hata, N., Barillot, C., Hornegger, J., Howe, R. (eds.) MICCAI 2014. LNCS, vol. 8675, pp. 305–312. Springer, Cham (2014). https://doi.org/10.1007/978-3-319-10443-0_39
2. Ben-Cohen, A., et al.: Cross-modality synthesis from CT to PET using FCN and GAN networks for improved automated lesion detection. Eng. Appl. Artif. Intell. **78**, 186–194 (2019)
3. Goodfellow, I., et al.: Generative adversarial nets. Advances in Neural Information Processing Systems, pp. 2672–2680 (2014)
4. Arslan, S., Ktena, S.I., Glocker, B., Rueckert, D.: Graph saliency maps through spectral convolutional networks: Application to sex classification with brain connectivity. arXiv preprint arXiv:1806.01764 (2018)
5. Ktena, S.I., et al.: Distance metric learning using graph convolutional networks: application to functional brain networks. In: Descoteaux, M., Maier-Hein, L., Franz, A., Jannin, P., Collins, D.L., Duchesne, S. (eds.) MICCAI 2017. LNCS, vol. 10433, pp. 469–477. Springer, Cham (2017). https://doi.org/10.1007/978-3-319-66182-7_54
6. Kipf, T.N., Welling, M.: Semi-supervised classification with graph convolutional networks. arXiv preprint arXiv:1609.02907 (2016)

7. Olut, S., Sahin, Y.H., Demir, U., Unal, G.: Generative adversarial training for MRA image synthesis using multi-contrast MRI. arXiv preprint arXiv:1804.04366 (2018)
8. Yang, H., et al.: Unpaired brain MR-to-CT synthesis using a structure-constrained CycleGAN. In: Stoyanov, D., et al. (eds.) DLMIA/ML-CDS -2018. LNCS, vol. 11045, pp. 174–182. Springer, Cham (2018). https://doi.org/10.1007/978-3-030-00889-5_20
9. Soussia, M., Rekik, I.: A review on image-and network-based brain data analysis techniques for Alzheimer's disease diagnosis reveals a gap in developing predictive methods for prognosis. arXiv preprint arXiv:1808.01951 (2018)
10. Pan, S., Hu, R., Long, G., Jiang, J., Yao, L., Zhang, C.: Adversarially regularized graph autoencoder. arXiv preprint arXiv:1802.04407 (2018)
11. Mahjoub, I., Mahjoub, M.A., Rekik, I.: Brain multiplexes reveal morphological connectional biomarkers fingerprinting late brain dementia states. Sci. Rep. **8**, 4103 (2018)
12. Lisowska, A., Rekik, I.: Alzheimer's disease neuroimaging initiative and others: joint pairing and structured mapping of convolutional brain morphological multiplexes for early dementia diagnosis. Brain Connect. **9**, 22–36 (2018)
13. Raeper, R., Lisowska, A., Rekik, I.: Cooperative correlational and discriminative ensemble classifier learning for early dementia diagnosis using morphological brain multiplexes. IEEE Access **6**, 43830–43839 (2018)
14. Soussia, M., Rekik, I.:Unsupervised manifold learning using high-order morphological brain networks derived from T1-w MRI for autism diagnosis. Front. Neuroinf. **12** (2018)
15. Wang, B., Ramazzotti, D., De Sano, L., Zhu, J., Pierson, E., Batzoglou, S.: SIMLR: a tool for large-scale single-cell analysis by multi-kernel learning. bioRxiv p. 118901 (2017)
16. Breiman, L.: Random forests. Mach. Learn. **45**, 5–32 (2001)
17. Fischl, B.: Freesurfer. Neuroimage **62**, 774–781 (2012)

Predicting High-Resolution Brain Networks Using Hierarchically Embedded and Aligned Multi-resolution Neighborhoods

Kübra Cengiz and Islem Rekik[✉]

BASIRA Lab, Faculty of Computer and Informatics,
Istanbul Technical University, Istanbul, Turkey
irekik@itu.edu.tr
http://basira-lab.com

Abstract. Several works have been dedicated to image super-resolution (i.e., synthesizing high-resolution data from low-resolution data). However, existing works only operate on images (e.g., predicting 7T-like magnetic resonance image (MRI) from 3T MRI) whereas brain connectivity network super-resolution remains unexplored. To fill this gap, we propose the first framework for predicting high-resolution (HR) brain networks from low-dimensional (LR) brain networks by hierarchically aligning and embedding LR neighborhood centered at the testing sample, along with its corresponding HR neighborhood. The proposed hierarchical embedding better preserves higher-order structural neighborhood of subjects within each domain. Recently, a seminal work was introduced for brain network prediction at a single resolution (or scale), where domain alignment was achieved using canonical correlation analysis followed by manifold learning to identify the most similar neighbors to the testing subject (i.e., testing neighborhood) in the source domain that can best predict the missing target network. Here, we inductively extend this idea by hierarchically learning the embedding and alignment of embedding of LR and HR neighborhoods. Our proposed framework achieved the best results in comparison with baseline methods.

1 Introduction

Neuroimaging studies associate autism spectrum disorder (ASD) with local structural and functional brain deficits [1,2]. Since ASD diagnosis is highly challenging, advanced machine learning-based diagnosis frameworks have been developed [3], most of which leveraging functional magnetic resonance imaging (fMRI) and diffusion tensor imaging (DTI) neuroimaging modalities [4–6]. Typically, there are two conventional representations of the brain derived from MRI data: (i) intensity images, and (ii) connectivity networks (also called connectome). To boost diagnosis accuracy of brain disorders, one would ideally use *high-resolution* brain images and connectomes. However, MRI data with high-resolution are very

© Springer Nature Switzerland AG 2019
I. Rekik et al. (Eds.): PRIME 2019, LNCS 11843, pp. 115–124, 2019.
https://doi.org/10.1007/978-3-030-32281-6_12

scarce due to the limited number of high-resolution 7T MRI scanners worldwide. To circumvent this issue, several works focused on designing methods for synthesizing high-resolution images (7T-like MR) from low-resolution images (3T MR) [7]. However, to the best of our knowledge, existing works on predicting high-resolution data (HR) from low-resolution (LR) data overlooked connectomic data, i.e., brain networks. Typically, a brain connectome is the result of time-consuming MRI data processing pipelines which integrate an image to brain atlas parcellation step such as Automated Anatomical Labelling (AAL) [8] with 90 anatomical regions of interest (ROIs), defining the resolution (or size) of the constructed brain connectome. To generate brain connectomes at different resolutions or scales, one generally needs to process and register the input MRI to each target MRI atlas space for automatic labelling of brain ROIs. However, the bench-to-bedside image processing pipeline to transform an MR image into a connectome is time-consuming –particularly when using high-resolution brain images and atlases. Alternatively, one can learn how to directly synthesize HR brain connectome from LR brain connectome to alleviate the computational cost of image processing including the major steps of registration and label propagation, which are highly prone to bias.

Interestingly, works on brain network to network prediction are very limited [9] with the exception of the recent work [10] proposing the first framework for missing multiple target brain networks prediction from single source brain network. Since multi-view brain networks have different distributions, [10] integrated a domain alignment step to find shared space where source and target networks are projected while maximizing the correlation between their respective distributions. Fundamentally, this work is based on predicting the missing target views from source views by learning how to select the best source training samples in the shared space in terms of (i) closeness to the testing subject in the LR domain, and (ii) their cross-domain overlap score based on the number of shared local neighbors these training samples have across source and target domains. We term the set of selected source training subjects, which are close to the testing subject, as the 'testing neighborhood' (TN). Next, the missing network in the target domain is predicted by linearly fusing the selected source training samples in the previous step. Although pioneering, this work is limited by the use of an inherently *flat* training sample selection, which overlooks the hierarchical structure that might be present in the testing neighborhood. In order to address these challenges, we propose *the first* framework that predicts a HR brain network from a LR brain network rooted in a hierarchical multi-layer embedding and alignment of LR and HR testing neighborhoods.

We base our method on a simple hypothesis: if one can identify *the best hierarchically embedded representations of neighborhood* including training samples centered around a given testing subject in the LR domain, one can use a weighted average of their corresponding samples in the HR domain to predict the missing testing HR network. To account for the domain shift where the distribution of the source LR and target HR domains might be misaligned, we first leverage canonical correlation analysis (CCA) to find a *coupled LR-HR manifold*

[11] that nests projected LR and HR networks while maximizing the correlation between their respective distributions. Next, we learn a subject-to-subject similarity matrix using multi-kernel connectomic manifold learning [12] which models the relationships between all training and testing samples in the coupled space. This defines the baseline layer of our HR prediction framework. Next, we propose to *hierarchically* learn and embed LR neighoborhoods centered at the LR testing sample and its corresponding HR neighborhood at each CCA-based domain alignment layer. In the last layer, we identify the most similar training samples *with the highest cross-domain scores* to the testing subject in the LR domain for prediction in the HR domain. Both domain alignment and manifold embedding steps are hierarchically implemented across L layers. Ultimately, we use the final output of aligned and embedded shared subspace to predict HR network. Specifically, we score the selected closest training samples to the testing LR network with the highest hierarchical cross-domain neighborhood overlap. We show that our proposed method achieves a better prediction accuracy in comparison with two baseline methods [10, 12].

2 Proposed Method

In this section, we detail our proposed hierarchical LR and HR domain alignment and testing neighborhood embedding for brain network super-resolution. We denote matrices by boldface capital letters, e.g., $\mathbf{X}$, and scalars by lowercase letters, e.g., x. We denote the transpose operator and the trace operator as $\mathbf{X}^T$ and $tr(\mathbf{X})$, respectively. We illustrate the important steps of the proposed pipeline in Fig. 1.

Feature Extraction. Each brain is represented by two of connectivity matrices in LR and HR domains, respectively (Fig. 1–A). Each element in a single matrix captures the relationship between two anatomical regions of interest (ROIs) using a specific metric (e.g., correlation between neural activity or similarity in brain morphology) and the number of ROIs of LR and HR are denoted n_1 and n, respectively. We then vectorize each connectivity matrix of the ith subject to define a feature vector $\mathbf{f}_{LR}^i$ (resp. $\mathbf{f}_{HR}^i$) for its HR (resp. LR) brain network. We concatenate the off-diagonal elements in the upper triangular part of the input matrices (LR and HR, respectively). Hence, each LR brain network is represented by $n_1 \times n_1$ matrix and the vectorization of the matrix produces a $(d_1 = n_1 \times (n_1 - 1)/2)$ size feature vector. Each HR network is encoded in $n \times n$ matrix which is vectorized into a $(d = n \times (n - 1)/2)$ size feature vector. Given N subjects, we leave-one-out cross-validation and store the remaining $(N - 1)$ training LR feature vectors in a training LR matrix $\mathbf{D}_{LR} \in \mathbb{R}^{(N-1) \times d_1}$ and HR feature vectors in training HR matrices $\mathbf{D}_{HR} \in \mathbb{R}^{(N-1) \times d}$.

    Step 1: CCA-based LR and HR domain alignment. Our main goal is to learn how to predict HR brain network from a given LR brain network (Fig. 1–B). However, this learning process might be sensitive to the domain fracture issue, where data distributions driven from different domains are not inherently and naturally aligned. To solve this issue and motivated by the fact

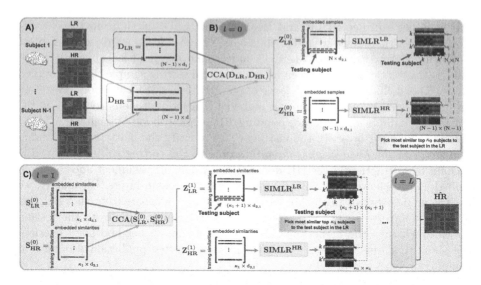

Fig. 1. *Pipeline of the proposed hierarchical multi-layer embedding and alignment of low-resolution (LR) and high-resolution (HR) neighborhoods for HR network prediction. (A) Each training subject has a LR network and a HR network. Each network is encoded in a symmetric connectivity matrix, whose upper off-diagonal part is vectorized. We store training LR feature vectors in a training LR matrix $\mathbf{D}_{LR}$ and HR feature vectors in training HR matrices $\mathbf{D}_{HR}$. (B) By using the training LR matrix $\mathbf{D}_{LR}$ and pairing it with its corresponding training HR matrix $\mathbf{D}_{HR}$, we learn a coupled LR-HR manifold using Canonical Correlation Analysis (CCA) for domain alignment. We then use multi-kernel learning [12] to learn a similarity matrix that models the relationship between training and testing subject embeddings $(\mathbf{Z}_{LR}^{(0)})$ in the coupled manifold in layer $l = 0$. We also learn a HR manifold that nests only embedded $(\mathbf{Z}_{HR}^{(0)})$ training subjects in the first layer. In the first layer, we primarily identify the top κ_0 training LR samples in the aligned domain with the highest learned similarities to the LR testing sample. This selected training set defines the testing neighborhood, which will be hierarchically mapped and aligned in the next layers. (C) We hierarchically learn a LR-HR domain alignment and neighborhood embedding and select a new set of top κ_l training subjects with the highest learned similarity scores to the embedded testing subject. In the last hierarchical layer ($l = L$), we select training hierarchically embedded LR samples which are (i) most similar to the hierarchically embedded LR testing samples, and (ii) have the highest cross-domain overlap in proximal neighbors. Last, we average the corresponding HR networks of the selected LR training networks in the last layer L to predict the target missing HR network.*

that canonical correlation analysis (CCA) is efficient in analyzing and mapping two sets of variables onto a shared aligned space [13,14], we learn CCA mappings that align LR brain networks with HR brain networks, respectively, to a shared space. Given a training LR matrix $\mathbf{D}_{LR} \in \mathbb{R}^{(N-1) \times d_1}$ comprising $N - 1$ training feature vectors, each of size d_1, and a training HR matrix $\mathbf{D}_{HR} \in \mathbb{R}^{(N-1) \times d}$, we estimate a LR mapping $\mathbf{W}_{LR}$ and a target mapping $\mathbf{W}_{HR}$ that transforms both onto the couple LR-HR space, respectively. This produces LR and HR embeddings ($\mathbf{Z}_{LR}^{(l)}$ and $\mathbf{Z}_{HR}^{(l)}$) in each layer l:

$$\mathbf{Z}_{LR}^{(l)} = \mathbf{D}_{LR}^{(l)}\mathbf{W}_{LR}^{(l)} \in \mathbb{R}^{(\kappa_l+1)\times d_{2,l}}$$
$$\mathbf{Z}_{HR}^{(l)} = \mathbf{D}_{HR}^{(l)}\mathbf{W}_{HR}^{(l)} \in \mathbb{R}^{\kappa_l\times d_{2,l}} \tag{1}$$

In the testing stage, we use the learned canonical transformation matrices to map the LR feature vector of a testing subject onto the shared space, where we *learn* how to identify the most *similar and trustworthy* training LR feature vectors to the testing LR network using LR and HR manifold learning.

Step 2: HR and LR manifold learning. Following the domain alignment, we learn pairwise similarities between LR samples (resp., HR) samples using the embeddings $\mathbf{Z}_{LR}^{(l)}$ (resp. $\mathbf{Z}_{HR}^{(l)}$) in the aligned coupled LR-HR space. Specifically, we leverage the recent work of [12] proposing to learn a convenient cell-to-cell similarity function from a single-cell data as an input. SIMLR (single-cell interpretation via multi-kernel learning) firstly clusters samples into groups for identification of subgroups and projects into low-dimensional. This method finds a best distance metric for fitting structure of different groups by combining multiple kernels. The main advantage of SIMLR is the flexibility of the adoption of multiple kernel representations for calculating similarities although single cell data have varied statistical characteristic. By using this method, we learn two manifolds: (1) one LR manifold encoded in $\mathbf{S}_{LR}$ similarity matrix $\mathbf{S}_{LR}$, which integrates all training and testing samples, and (2) one HR manifold encoded in $\mathbf{S}_{HR}$, modeling the relationship between training HR samples. (Fig. 1–B). Each kernel $\mathbf{K}$ is Gaussian and expressed as follows: $\mathbf{K}(\mathbf{f}_{LR}^i, \mathbf{f}_{LR}^j) = \frac{1}{\epsilon_{ij}\sqrt{2\pi}}e^{\left(-\frac{|\mathbf{f}_{LR}^i-\mathbf{f}_{LR}^j|^2}{2\epsilon_{ij}^2}\right)}$, where $\mathbf{f}_{LR}^i$ and $\mathbf{f}_{LR}^j$ denote the feature vectors of the i-th and j-th subjects respectively and ϵ_{ij} is defined as: $\epsilon_{ij} = \sigma(\mu_i + \mu_j)/2$, where σ is a tuning parameter and $\mu_i = \frac{\sum_{l\in KNN(\mathbf{f}_{LR}^i)}|\mathbf{f}_{LR}^i-\mathbf{f}_{LR}^j|}{k}$, where $KNN(\mathbf{f}_{LR}^i)$ represents the top neighboring subjects i of subject j. The learned similarity matrices in both LR and HR *aligned* domains should be small if the distance between a pair of subjects is large.

$$\mathbf{S}_{LR}^{(l)} = SIMLR(\mathbf{Z}_{LR}^{(l)}) \in \mathbb{R}^{(\kappa_l+1)\times d_{3,l}}$$
$$\mathbf{S}_{HR}^{(l)} = SIMLR(\mathbf{Z}_{HR}^{(l)}) \in \mathbb{R}^{\kappa_l\times d_{3,l}} \tag{2}$$

For simplicity, in the following sections we will abstract away the internal structure of the SIMLR in Eq. 2 and use $\mathbf{S}_{LR}^{(l)}$ and $\mathbf{S}_{HR}^{(l)}$ to denote an arbitrary SIMLR module learning the new embedding of similarities in aligned LR and HR domains, respectively.

Step 3: HR network prediction via hierarchical alignment and embedding of LR and HR testing neighborhoods. Our proposed HR prediction framework hierarchically learns a finer LR testing neighborhood embedding along with its corresponding HR neighborhood. The key idea is to learn the most similar top κ subjects to the testing subject in LR domain at layer l, by using the embeddings of neighbors generated from the previous layer $l-1$. Next, both learned LR and HR neighborhood embeddings are aligned using CCA (Fig. 1–C). Given the baseline learned similarity matrix $\mathbf{S}_{LR}^{(0)}$ at layer $l = 0$, we detail below how the hierarchical alignment and embedding modules of $\mathbf{S}_{LR}^{(0)}$ operate and how this process is iterated from layer l to layer $l+1$.

• *Hierarchical alignment and embedding module.* At baseline layer $l = 0$, we leverage multiple kernel manifold learning [12] (**Step 2**) to learn (i) the similarity matrix $\mathbf{S}_{LR}^{(0)} \in \mathbb{R}^{(\kappa_1 \times d_{3,1})}$ where $d_{3,1} < d_{2,1}$ between the testing LR network and training LR networks, and (ii) the similarity matrix $\mathbf{S}_{HR}^{(0)} \in \mathbb{R}^{(\kappa_1 \times d_{3,1})}$ between all training HR networks.

$\mathbf{S}_{LR}^{(l)} \in \mathbb{R}^{(\kappa_l+1) \times (\kappa_l+1)}$ denotes the most similar embedded LR samples to the embedded testing subject by SIMLR (i.e., embedding testing neighborhood) and $\mathbf{S}_{HR}^{(l)} \in \mathbb{R}^{\kappa_l \times \kappa_l}$ the corresponding embedded HR samples in layer l. κ_l denotes the dimension of the embedded neighborhood in layer l.

Suppose that $\mathbf{S}^{(l)}$ has already been computed, i.e., that we have computed the matrix in the l-th layer of our model. Given this input, the hierarchical $l+1$ layer generates a new similarity matrix of training LR embeddings $\mathbf{Z}_{LR}^{(l+1)}$ and $\mathbf{Z}_{HR}^{(l+1)}$. In particular, we *alternatingly* apply the two following equations:

$$\mathbf{Z}_{LR}^{(l+1)} = CCA^{(l)}(\mathbf{S}_{LR}^{(l)}) \in \mathbb{R}^{(\kappa_l+1) \times d_{2,l}}$$
$$\mathbf{Z}_{HR}^{(l+1)} = CCA^{(l)}(\mathbf{S}_{HR}^{(l)}) \in \mathbb{R}^{\kappa_l \times d_{2,l}} \tag{3}$$

$$\mathbf{S}_{LR}^{(l+1)} = SIMLR^{(l)}(\mathbf{Z}_{LR}^{(l+1)}) \in \mathbb{R}^{(\kappa_l+1) \times (\kappa_l+1)}$$
$$\mathbf{S}_{HR}^{(l+1)} = SIMLR^{(l)}(\mathbf{Z}_{HR}^{(l+1)}) \in \mathbb{R}^{\kappa_l \times \kappa_l} \tag{4}$$

Step 4: Predicting HR networks using cross-domain shared neighborhood. Finally, in the last layer L, once the the most similar LR training neighbors to the testing LR sample with the highest cross-domain scores are identified, we retrieve their corresponding networks in the HR domain, then use weighted average to predict the target missing HR network. Basically, we define a 'trust score' for each training sample i similar to the testing subject j based on the overlap of their hierarchically embedded neighborhoods in the aligned LR and HR domains, respectively. Following the learning of $\mathbf{S}_{LR}$ using all samples in the mapped source domain using SIMLR, we identify the top κ-closest training subjects to a given testing subject. Next, for each training sample, we find its nearest neighbors using $\mathbf{S}_{LR}$ and $\mathbf{S}_{HR}$, learned in the aligned target domain using only training subjects (Fig. 1).

This is rooted in the assumption that for a particular training subject which is close to the testing subject in the LR domain, the more shared neighbors it has across the embedded LR and HR neighborhoods in the last layer L, the more reliable it is in predicting the HR from the LR network, and thus it can be considered as trustworthy for the target prediction task. We compute a normalized trust score (TS) for each closest training neighbor to the testing subject by (i) first identifying the list of its top κ closest neighbors $\mathcal{N}_{LR}$ in $\mathbf{S}_{LR}$ and $\mathcal{N}_{HR}$ in $\mathbf{S}_{HR}$, then (ii) computing the normalized overlap between both lists as $TS(\kappa) = \frac{\mathcal{N}_{LR} \bigcap \mathcal{N}_{HR}}{\kappa}$. The ultimate $TN(\kappa)$ score is thus calculated as a soft overlap between $\mathbf{S}_{LR}^{(L)}$ and $\mathbf{S}_{HR}^{(L)}$ weighted by $\mathbf{S}_{LR}$.

3   Results and Discussion

Connectomic Dataset and Method Parameters. We used leave-one-out cross-validation to evaluate the proposed prediction framework on 186 normal controls (NC) from Autism Brain Imaging Data Exchange (ABIDE I)[1] public dataset, each with structural T1w MR image. We used FreeSurfer [15] to reconstruct both right and left cortical hemispheres for each subject from T1-w MRI, and then parcellated each cortical hemisphere into 35 cortical regions using Desikan-Killiany Atlas. For each subject, we created cortical morphological brain networks derived from the cortical maximum principal curvature using the technique proposed in [16–18]. For SIMLR, we used a nested grid search, fixing the number of clusters c ($1 \leq c \leq 5$). We used 10 kernels. We set the number of $L = 3$ and the number of selected neighbors in each layer is defined as $\kappa = \{50, 25, 5\}$, respectively.

LR Data Synthesis via Downsampling HR Brain Connectomes. HR data downsampling or degradation models are frequently used in *image* super-resolution literature (i.e, MR images) for evaluation. For instance, [19] applied downsampling method to obtain LR images of 256 × 256 and 128 × 128 resolutions from 512 × 512 HR images. By doing so, downsampling decreases the number of voxels and also causes the loss of image details, thereby creating a lower-resolution image. Similarly, we create LR networks for each subject by computing mean connectivity value of HR within a $w \times w$ sized window, where w denotes the window size ($w = 10, 16, 20$). Hence, we created three different LR network datasets through this mean-pooling process.

Evaluation and Comparison Methods. To evaluate the performance of our hierarchical HR prediction from LR framework, we benchmark our framework against: (1) [12] where we used SIMLR to identify the most similar neighbors to the left-out testing subject the LR domain without any domain alignment, (2) the baseline network prediction method integrating both manifold learning and aligned proposed by [10]. (Figure 2–A–C) demonstrates that our method achieves the lowest prediction mean absolute error (MAE) in comparison with baseline methods. Figure 2–D displays the predicted HR networks by different methods from an input LR network along with the residual networks in both

---

[1] http://fcon_1000.projects.nitrc.org/indi/abide/.

left and the right hemispheres for a representative testing subject. Clearly, our method decreases the residual error. However, we would like to point out that our framework training is constrained by the availability of *paired* LR and HR training networks. In our future work, we will relax this constraint by allowing our high-resolution prediction framework to learn from *unpaired* training LR and HR brain connectomes.

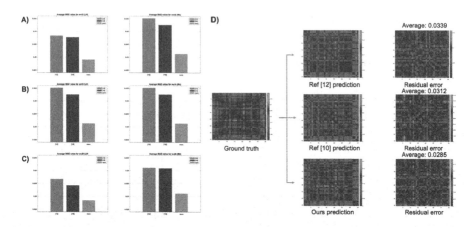

Fig. 2. (A–C)*Evaluating the prediction performance of our proposed hierarchical alignment and embedding of LR and HR neighborhoods on left and right hemispheric brain networks (LH and RH).* We report the mean absolute error (MAE) between ground-truth and predicted HR networks. We benchmark agains two methods: (1) [12] where we used SIMLR to identify the most similar neighbors to the left-out testing subject the LR domain without any domain alignment, (2) the baseline network prediction method integrating both manifold learning and aligned proposed by [10]. (D) *Comparison between the ground-truth and predicted HR networks from LR networks (obtained by mean-pooling using a w = 10 sized window) of the left hemisphere for a representative testing subject by our method and comparison methods.* We display the residual matrices computed using element-wise absolute difference between ground truth and predicted networks. Ground truth: the ground truth HR network of a testing subject. Prediction: the predicted HR network using our purposed framework.

4   Conclusion

This paper proposes the first work on predicting high-resolution brain networks from low-resolution brain networks, by bridging the connection between low-resolution and high-resolution domains, then hierarchically learning how to create high-order nested neighborhood embeddings for ultimately identifying the most reliable training LR samples for the target prediction task. In our future work, we will learn how to predict a multi-resolution brain networks from a single low-resolution brain network. We will also evaluate our hierarchical HR prediction framework on larger datasets to predict other types of high-resolution brain networks including functional brain connectivity and structural connectivity.

References

1. Hollander, E., et al.: Striatal volume on magnetic resonance imaging and repetitive behaviors in autism. Biol. Psychiatry **58**, 226–232 (2005)
2. Rojas, D.C., Smith, J.A., Benkers, T.L., Camou, S.L., Reite, M.L., Rogers, S.J.: Hippocampus and amygdala volumes in parents of children with autistic disorder. Am. J. Psychiatry **161**, 2038–2044 (2004)
3. Hyde, K.K., et al.: Applications of supervised machine learning in autism spectrum disorder research: a review. Rev. J. Autism Dev. Disord. **6**, 128–146 (2019)
4. Rane, P., Cochran, D., Hodge, S.M., Haselgrove, C., Kennedy, D., Frazier, J.A.: Connectivity in autism: a review of MRI connectivity studies. Harv. Rev. Psychiatry **23**, 223 (2015)
5. Koshino, H., Carpenter, P.A., Minshew, N.J., Cherkassky, V.L., Keller, T.A., Just, M.A.: Functional connectivity in an fMRI working memory task in high-functioning autism. Neuroimage **24**, 810–821 (2005)
6. Price, T., Wee, C.-Y., Gao, W., Shen, D.: Multiple-network classification of childhood autism using functional connectivity dynamics. In: Golland, P., Hata, N., Barillot, C., Hornegger, J., Howe, R. (eds.) MICCAI 2014. LNCS, vol. 8675, pp. 177–184. Springer, Cham (2014). https://doi.org/10.1007/978-3-319-10443-0_23
7. Bahrami, K., Shi, F., Rekik, I., Shen, D.: Convolutional neural network for reconstruction of 7T-like images from 3T MRI using appearance and anatomical features. In: Carneiro, G., et al. (eds.) LABELS/DLMIA -2016. LNCS, vol. 10008, pp. 39–47. Springer, Cham (2016). https://doi.org/10.1007/978-3-319-46976-8_5
8. Tzourio-Mazoyer, N., et al.: Automated anatomical labeling of activations in SPM using a macroscopic anatomical parcellation of the MNI MRI single-subject brain. Neuroimage **15**, 273–289 (2002)
9. Soussia, M., Rekik, I.: A review on image-and network-based brain data analysis techniques for Alzheimer's disease diagnosis reveals a gap in developing predictive methods for prognosis. arXiv preprint arXiv:1808.01951 (2018)
10. Zhu, M., Rekik, I.: Multi-view brain network prediction from a source view using sample selection via CCA-based multi-kernel connectomic manifold learning. In: Rekik, I., Unal, G., Adeli, E., Park, S.H. (eds.) PRIME 2018. LNCS, vol. 11121, pp. 94–102. Springer, Cham (2018). https://doi.org/10.1007/978-3-030-00320-3_12
11. Blitzer, J., Kakade, S., Foster, D.: Domain adaptation with coupled subspaces. In: Proceedings of the Fourteenth International Conference on Artificial Intelligence and Statistics, pp. 173–181 (2011)
12. Wang, B., Zhu, J., Pierson, E., Ramazzotti, D., Batzoglou, S.: Visualization and analysis of single-cell rna-seq data by kernel-based similarity learning. Nature Methods **14**, 414 (2017)
13. Zhu, X., Suk, H.I., Lee, S.W., Shen, D.: Canonical feature selection for joint regression and multi-class identification in Alzheimer's disease diagnosis. Brain Imaging Behav. **10**, 818–828 (2016)
14. Haghighat, M., Abdel-Mottaleb, M., Alhalabi, W.: Fully automatic face normalization and single sample face recognition in unconstrained environments. Expert Syst. Appl. **47**, 23–34 (2016)
15. Fischl, B.: FreeSurfer. Neuroimage **62**, 774–781 (2012)
16. Mahjoub, I., Mahjoub, M.A., Rekik, I.: Brain multiplexes reveal morphological connectional biomarkers fingerprinting late brain dementia states. Sci. Rep. **8**, 4103 (2018)

17. Soussia, M., Rekik, I.: Unsupervised manifold learning using high-order morpho-
    logical brain networks derived from T1-w MRI for autism diagnosis. Front. Neu-
    roinform. **12** (2018)
18. Nebli, A., Rekik, I.: Gender differences in cortical morphological networks. Brain
    Imaging Behav. 1–9 (2019)
19. Wang, Y.H., Qiao, J., Li, J.B., Fu, P., Chu, S.C., Roddick, J.F.: Sparse
    representation-based MRI super-resolution reconstruction. Measurement **47**, 946–
    953 (2014)

Catheter Synthesis in X-Ray Fluoroscopy with Generative Adversarial Networks

Ihsan Ullah, Philip Chikontwe, and Sang Hyun Park[✉]

Department of Robotics Engineering, DGIST, Daegu 42988, Republic of Korea
{ihsankhan,philipchicco,shpark13135}@dgist.ac.kr

Abstract. Accurate localization of catheters or guidewires in fluoroscopy images is important to improve the stability of intervention procedures as well as the development of surgical navigation systems. Recently, deep learning methods have been proposed to improve performance, however these techniques require extensive pixel-wise annotations. Moreover, the human annotation effort is equally expensive. In this study, we mitigate this labeling effort using generative adversarial networks (cycleGAN) wherein we synthesize realistic catheters in flouroscopy from localized guidewires in camera images whose annotations are cheaper to acquire. Our approach is motivated by the fact that catheters are tubular structures with varying profiles, thus given a guidewire in a camera image, we can obtain the centerline that follows the profile of a catheter in an X-ray image and create plausible X-ray images composited with such a centerline. In order to generate an image similar to the actual X-ray image, we propose a loss term that includes perceptual loss alongside the standard cycle loss. Experimental results show that the proposed method has better performance than the conventional GAN and generates images with consistent quality. Further, we provide evidence to the development of methods that leverage such synthetic composite images in supervised settings.

Keywords: Adversarial learning · Catheter robot · Convolutional neural networks · Image translation · Image synthesis

1 Introduction

Catheterization is an important process extensively applied during interventional therapy in which catheters are inserted into the body for disease treatment or surgical procedures offering low risk for patients and quick recovery. Catheters are tailored for different forms of intervention such as cardiovascular, neurovascular and gastrointestinal wherein several tasks can be performed depending on the type of catheter. In cardiac catherterization, expert physicians insert a guidewire into an artery or vein and transport stent under fluoroscopic guidance following placement. However, in most cases catheters are thin flexible tubes with varied stiffness which can be easily confused with anatomy during placement

---

Equally contributed by Mr. Ullah and Mr. Chikontwe.

© Springer Nature Switzerland AG 2019
I. Rekik et al. (Eds.): PRIME 2019, LNCS 11843, pp. 125–133, 2019.
https://doi.org/10.1007/978-3-030-32281-6_13

and may lead to malpositioning with serious complications. Thus, conventional catheterization requires prolonged treatment and many contrast medications in high concentration for improved visibility.

To alleviate the potential for injury and radiation exposure to experts and patients, robot catheter systems are being developed; however, precise and reliable detection of catheters in these systems is a challenging task. Accordingly, image based catheter detection and tracking methods have been proposed. For example, the Frangi feature introduced with supervised learning showed successful detection of instruments [5]. Later, rather than single features, improved features such as segment-like features (SEGlet) were proposed to address robust catheter tracking in video sequences under large deformations [6]. However, these methods require domain expertise and do not guarantee accurate results especially in noisy environments.

Recently, deep learning based methods have achieved promising results [3, 7,8] with improved classification and feature extraction capabilities. However, the application of deep learning methods for catheter localization is hindered mainly by the lack of significant big data and expensive data annotation. To address this problem, Yi et al. [8] proposed simulated catheter data generation and applied a multi-scale approach via recurrent modules along with iterative refinement for catheter and tube detection in pediatric X-ray images. However, due to large domain shift between simulated training and test images; detection fails in some cases especially when catheters are projected over the abdomen.

Recently, generative adversarial networks (GAN) have shown good performance in generating realistic X-ray or CT images. For example, Tmenova et al. [4] propose a method for anatomy-based data augmentation in X-ray angiography using CycleGAN [12] for images containing both low and high contrast. A CT based motion simulator was employed to generate artifical X-ray with varying projected vessel types. However, the generated images contain distorted textures far from realistic anatomy and requires detailed parameter tuning to simulate varied vessels. More recently, X2CT-GAN [9] was introduced to address the task of synthetic X-ray generation from 3D body surface meshes. The authors demonstrate the potential of the method in interventional procedures such as anomaly detection and X-ray completion from partial views. Moreover, a CycleGAN based style transfer method is employed to learn mapping from 2D to 3D.

In this work, we address the task of cardiac X-ray image synthesis which is applicable for data-augmentation of X-ray images containing inserted catheters. To generate realistic catheters, we first extract the centerline of a guidewire contained in a camera image which is relatively easy to acquire and process, then composite the centerline with a real X-ray image. Simply compositing the extracted guidewire centerline with real X-ray images produces visually unnatural images. To address this, we leverage the cycleGAN based method to generate the realistic X-ray images from the initial composite X-ray. We propose a loss function that combines the standard cycle loss with perceptual loss for improved visual quality. Perceptual loss is employed as a feature based optimization rather than using a pixel-based optimization only in the form of cycle loss. Experimental results demonstrate the advantage of the proposed loss both visually and quantitatively.

2   Methods

This work addresses the synthesis of catheters in X-ray from guidewires in natural images used in interventional procedures. In this section, the datasets, neural network architectures, and objective functions employed are described.

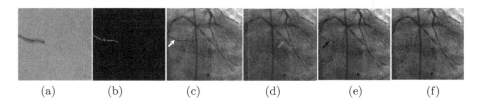

| (a) | (b) | (c) | (d) | (e) | (f) |

Fig. 1. (a) Guidewire in natural image (b) centerline using [10] (c) composited X-ray with guidewire centerline shown with *red arrow* (d) X-ray image with guidewire highlighted with *yellow arrow* (e) composited X-ray with guidewire centerline highlighted with *green arrow* (f) reconstructed X-ray. (Color figure online)

Datasets.[1] Figure 1 shows sample images of the two main datasets used in this work. The first set (Fig. 1(a)) consists 11 videos containing a total 11,884 2D sequences of guidewires exhibiting varied movement in different directions with respect to magnetic field changes in eight coils of a robotic system. It was collected at our inhouse microrobot research centre. The second (Fig. 1(f)), is a set of 2D coronary angiography X-rays of 100 patients containing left and right corononary angiograms obtained with several cranial and caudal views i.e, Left Anterior Oblique (LAO) cranial, Anterior-Posterior (AP) cranial, Anterior-Posterior (AP) caudal and so on.

Formulation. Our goal is to learn mapping functions between two domains using GANs. In the classical adversarial setting a two player mini-max game between a generator G and a discriminator network D is played, wherein G is trained to map random vectors $z \in \mathcal{R}^z$ to a synthetic vector $a = G(z)$, with D distinguishing real samples from synthetic samples. We consider the task of synthesizing catheters in X-ray as an unsupervised image-to-image translation task. Thus, we employ the CycleGAN [12] architecture and explore the use of different objectives to achieve realistic synthesis. Figure 2 shows the framework.

Let X be the domain of X-ray images composited with guidewire centrelines $x_i \in X^n$ and Y be the domain of the original X-ray images $y_i \in Y^n$ with no existing catheter annotations. Given a natural image containing a guidewire $c_i \in \mathcal{C}^n$, we can obtain a composite $x_i = \hat{c}_i \oplus y_i$ where $\hat{c}_i = f(c_i)$ is the binary mask containing the centerline obtained by skeletonization method f [10].

---

[1] Note that no annotations are present for the X-ray dataset. Figure 1(d) only highlights the guidewire position for clarity.

We learn the mapping between $G : X \to Y$ and $F : Y \to X$ parametrized by the generators G_X, G_Y and adversarial discriminators D_X and D_Y, respectively. For the generators, we follow a mix between U-Net and residual (ResNet) style architectures with skip connections from the initial convolutions to the upsampled bottleneck layers including leakyReLU, instance normalization and tanh activations at the final layer. PatchGAN based discriminator [1] is used for D_X and D_Y with $K \times K$ receptive fields. Formally, the objective function of $G : X \to Y$ and D_Y is expressed as:

$$min_G max_{D_Y} \mathcal{L}_{adv}(G, D_Y, X, Y) = \mathbb{E}[log D_Y(y)] + \mathbb{E}[log(1 - D_Y(G(x)))], \quad (1)$$

to create generated images $G(x)$ similar to domain Y. Similarly for $F : Y \to X$ the objective is $min_F max_{D_X} \mathcal{L}_{adv}(F, D_X, Y, X)$. In addition, cycle-consistency between the generated samples in both the forward $(X \to Y)$ and backward $(Y \to X)$ cycles is achieved using

$$\mathcal{L}_{cyc}(G, F) = \mathbb{E}[||F(G(x)) - x||_1] + \mathbb{E}[||G(F(y) - y)||_1], \quad (2)$$

with the final objective being

$$\mathcal{L}(G, F, D_X, D_Y) = \mathcal{L}_{adv}(G, D_Y, X, Y) + \mathcal{L}_{adv}(F, D_X, Y, X) + \lambda \mathcal{L}_{cyc}(G, F). \quad (3)$$

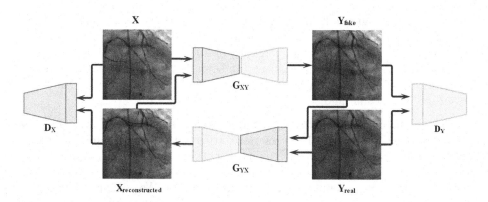

Fig. 2. Overall framework for catheter synthesis in a X-ray image

In addition to Eq. (3), we explore the use of structural similarity index (SSIM) [11] loss as well as perceptual loss [2]. The motivation for using SSIM is to enable the network to produce visually consistent images, formally SSIM for a pixel p is defined as

$$SSIM(p) = \frac{2\mu_x\mu_y + C_1}{\mu_x^2 + \mu_y^2 + C_1} \cdot \frac{2\sigma_{xy} + C_2}{\sigma_x^2 + \sigma_y^2 + C_2}, \mathcal{L}_{ssim}(P) = \frac{1}{N}\sum_{p \in P} 1 - SSIM(p). \quad (4)$$

with μ and σ representing mean and standard deviation of the inputs computed using a gaussian filter. $\mathcal{L}_{ssim}$ is computed between the forward and backward cycles i.e, $\mathcal{L}_{ssim}(\mathcal{L}_{cyc}G, x)$ and $\mathcal{L}_{ssim}(\mathcal{L}_{cyc}F, y)$.

On the other hand, perceptual loss $\mathcal{L}_{perp}$ has been reported to be more effective than per-pixel losses by depending on high-level features from pre-trained networks than depending on low-level pixel information. In this work, to incorporate this loss we use features from an ImageNet pre-trained ResNet by minimizing the loss between both the forward and backward cycles in a similar fashion with SSIM.

3 Experiments

3.1 Experimental Setup

Our methods are implemented using the Keras library with a tensorflow backend. We trained the cycleGAN networks using different generators i.e, ResNet and U-Net on the dataset created using composited images with publicly available implementations of the described networks. Notably, X-ray angiography is known for depicting different artifacts, thus we manually selected clean samples with little to no artifacts as well as images that contain the inserted guidewires for training and testing i.e, 1000 training, 500 validation and 1000 for test. Images are resized to 256×256 and trained with a batch size of 8 for 50 epochs producing output images of the same size. We used Adam optimizer with a learning rate of $2e - 4$ and fixed the trade-off co-efficient in Eq. (3) to 10 ($\lambda = 10$). During training, we randomly select centerline masks (obtained offline) for any given X-ray image as a form of data augmentation including random flips.

Evaluation of generative models is often challenging, especially when ground truths are not implied in the experimental setting. So, to analyze performance we employ visual judgments including standard qualitative metrics i.e, $\mathcal{L}_1$ error, SSIM and peak signal-to-noise ratio (PSNR). We evaluated these measures between the generated images $G_X(x)$ and the original X-ray images $y_i \in Y^n$. Furthermore, we present quantitative results of training the generated images in a supervised setting of catheter segmentation with the centerline from the natural camera image as ground-truth. This extra setup serves to give insight as to whether we can indeed use the labels obtained in natural guidewire images to detect the true catheter in the X-ray for which no labels are present.

3.2 Experimental Results

Qualitative performance of different models is compared in Table 1. CycleGAN with ResNet generators G_X, G_Y trained with $\mathcal{L}_{perp}$ loss obtained the highest PSNR score; implying the model generated samples with the least noise. Also, the lowest $\mathcal{L}_1$ error of 0.34 is obtained by this model verifying the benefit of perceptual loss in the training objectives despite the baseline model trained using only the regular cycle loss achieving the lowest SSIM error. Further, the

Table 1. Comparison of the models with different loss functions with respect to the average PSNR, SSIM and $\mathcal{L}_1$ error on the same test set. For PSNR, higher is better. Lower is better for SSIM and $\mathcal{L}_1$.

Model	(G_X, G_Y)	Loss	PSNR (dB)	SSIM error	$\mathcal{L}_1$ loss
CycleGAN [12]	ResNet	$\mathcal{L}_{cyc}$	7.2438	**0.0296**	0.4161
	U-Net	$\mathcal{L}_{cyc}$	7.5837	0.0400	0.3917
	ResNet	$\mathcal{L}_{cyc} + \mathcal{L}_{perp}$	**8.2862**	0.0570	**0.3404**
	U-Net	$\mathcal{L}_{cyc} + \mathcal{L}_{perp}$	7.8868	0.0624	0.3581
	ResNet	$\mathcal{L}_{cyc} + \mathcal{L}_{ssim}$	5.4591	0.0532	0.4726
	U-Net	$\mathcal{L}_{cyc} + \mathcal{L}_{ssim}$	7.6273	0.0556	0.3718

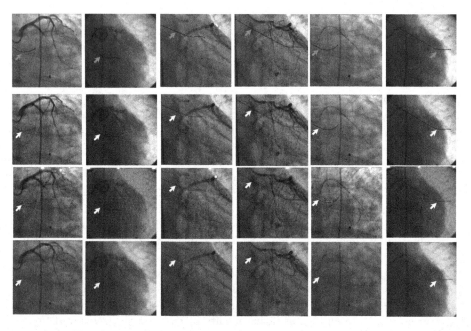

Fig. 3. From top to bottom: inputs, U-Net $\mathcal{L}_{cyc}$, ResNet $\mathcal{L}_{cyc}$ and ResNet $\mathcal{L}_{cyc} + \mathcal{L}_{perp}$. (Red arrows) initial position of the composited catheter and (green arrows) indicate the position of catheter in the generated images using different methods. (Color figure online)

model trained with both regular cyclegan loss and perceptual performs on par the best model i.e, both PSNR and $\mathcal{L}_1$ are second only to the best ResNet based generator method.

In Fig. 3, the qualitative results of the best model are illustrated. Notably, given the composited X-ray image as input, the generator outputs samples with the guidewire blended in the X-ray as though it were originally part of the image. More importantly, the output image maintains the visual quality of the

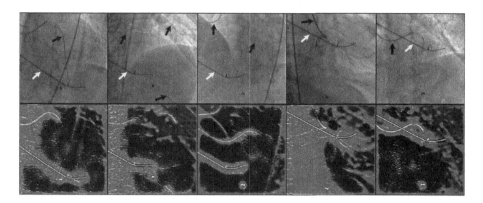

Fig. 4. Intermediate training results of a segmentation task using composite images. From top to bottom: input and probability map output. (yellow arrows) Inpainted catheter centerline from guidewire in a natural image, (red arrows) highlighted true inserted catheter. (Color figure online)

input image with a focus on the catheter region though no explicit supervision is provided. However, the generated samples using U-Net with $\mathcal{L}_{cyc}$ (2nd row) appear darker with no big change compared to the input images. On the other hand, the generated samples using ResNet with $\mathcal{L}_{cyc}$ (3rd row) only, introduced some artifacts on the catheter positions(green-arrows), while the samples generated using the proposed ResNet with $\mathcal{L}_{cyc} + \mathcal{L}_{perp}$ (4th row) shows well blended catheters with improved visual quality. We note that we empirically choose to train the models for a limited number of epochs; as training the models for long epochs may result in outputs with no visual indication of the initial composited catheter (Fig. 3 red arrows) i.e, the model will learn to generate images without the catheter. Moreover, given the initial input images we surmise an expert may be able to easily flag the catheter (centerline region) as artificial, opposed to the generated images whose inpainted catheter may be indistinguishable from the existing inserted catheters or mistaken for vessels in some cases.

Synthetic Samples for Supervised Settings. In Fig. 4, we show results of training a U-Net segmentation model with centerlines as ground-truth. Notably, during training the network is able to segment regions similar to the centerline including the true catheter whose label is we did not provide. This provides evidence that centerlines can aid in the detection of the true catheter. Training the network until convergence produces a model that ignores the true catheter and only accurately segments the centerline. For this task, we report 84.42%, 85.73% and 84.69% for the Dice, Precision and Recall scores, respectively. Achieving optimal results is not the goal of this task; only to show we can indeed use such weak labels to potentially segment the true catheter.

4   Discussion and Conclusion

In this work, we explored the possibility of using easily obtainable annotations of guidewires in natural images to synthesize catheters in X-ray images. To solve this task, we considered catheter synthesis as an unsupervised image-to-image translation problem in the generative adversarial framework. Our experiments demonstrate that perceptual losses produce high quality and less noisy images. Moreover, our approach can be used for obtaining more training data to train supervised methods with the possibility to generate annotations of catheters in X-ray whose labeling is expensive. In future works, we will introduce methods for both weakly-supervised detection and segmentation of catheters in different domains.

Acknowledgment. This work is supported by the Robot industry fusion core technology development project through the Korea Evaluation Institute of Industrial Technology (KEIT) funded by the Ministry of Trade, Industry and Energy of Korea (MOTIE) (NO. 10052980) and the DGIST R & D Program of the Ministry of Science and ICT (19-RT-01).

References

1. Isola, P., Zhu, J.Y., Zhou, T., Efros, A.A.: Image-to-image translation with conditional adversarial networks. In: The IEEE Conference on Computer Vision and Pattern Recognition (CVPR), July 2017
2. Johnson, J., Alahi, A., Fei-Fei, L.: Perceptual losses for real-time style transfer and super-resolution. In: Leibe, B., Matas, J., Sebe, N., Welling, M. (eds.) ECCV 2016. LNCS, vol. 9906, pp. 694–711. Springer, Cham (2016). https://doi.org/10.1007/978-3-319-46475-6_43
3. Subramanian, V., Wang, H., Wu, J.T., Wong, K.C., Sharma, A., Syeda-Mahmood, T.: Automated detection and type classification of central venous catheters in chest X-rays. arXiv preprint arXiv:1907.01656 (2019)
4. Tmenova, O., Martin, R., Duong, L.: CycleGAN for style transfer in X-ray angiography. Int. J. Comput. Assist. Radiol. Surg. 1–10 (2019)
5. Uherčík, M., Kybic, J., Zhao, Y., Cachard, C., Liebgott, H.: Line filtering for surgical tool localization in 3D ultrasound images. Comput. Biol. Med. **43**(12), 2036–2045 (2013)
6. Vandini, A., Glocker, B., Hamady, M., Yang, G.Z.: Robust guidewire tracking under large deformations combining segment-like features (SEGlets). Med. Image Anal. **38**, 150–164 (2017)
7. Wagner, M.G., Laeseke, P., Speidel, M.A.: Deep learning based guidewire segmentation in X-ray images. In: Medical Imaging 2019: Physics of Medical Imaging. vol. 10948, p. 1094844. International Society for Optics and Photonics (2019)
8. Yi, X., Adams, S., Babyn, P., Elnajmi, A.: Automatic catheter and tube detection in pediatric X-ray images using a scale-recurrent network and synthetic data. J. Digit. Imaging 1–10 (2019)
9. Ying, X., Guo, H., Ma, K., Wu, J., Weng, Z., Zheng, Y.: X2CT-GAN: reconstructing CT from biplanar X-rays with generative adversarial networks. In: Proceedings of the IEEE Conference on Computer Vision and Pattern Recognition, pp. 10619–10628 (2019)

10. Zhang, T., Suen, C.Y.: A fast parallel algorithm for thinning digital patterns. Commun. ACM **27**(3), 236–239 (1984)
11. Zhao, H., Gallo, O., Frosio, I., Kautz, J.: Loss functions for image restoration with neural networks. IEEE Trans. Comput. Imaging **3**(1), 47–57 (2016)
12. Zhu, J.Y., Park, T., Isola, P., Efros, A.A.: Unpaired image-to-image translation using cycle-consistent adversarial networks. In: Proceedings of the IEEE International Conference on Computer Vision, pp. 2223–2232 (2017)

Prediction of Clinical Scores for Subjective Cognitive Decline and Mild Cognitive Impairment

Aojie Li[1], Ling Yue[3,4(✉)], Manhua Liu[1,2(✉)], and Shifu Xiao[3,4(✉)]

[1] Department of Instrument Science and Engineering, School of EIEE,
Shanghai Jiao Tong University, Shanghai 200240, China
mhliu@sjtu.edu.cn
[2] MoE Key Lab of Artificial Intelligence, AI Institute,
Shanghai Jiao Tong University, Shanghai, China
[3] Department of Geriatric Psychiatry, Shanghai Mental Health Center,
Shanghai Jiao Tong University School of Medicine, Shanghai, China
bellinthemoon@hotmail.com, xiaoshifu@msn.com
[4] Alzheimer's Disease and Related Disorders Center,
Shanghai Jiao Tong University, Shanghai, China

Abstract. Mild cognitive impairment (MCI) is a neurological disorder that occurs in older adults involving cognitive impairments. It may occur as a transitional stage between normal aging and dementia such as Alzheimer's disease (AD). Recent studies found that subjective cognitive decline (SCD) may be the early clinical precursor of dementia that precedes MCI. SCD individuals with normal cognition may already have some medial temporal lobe atrophy. This paper proposes a machine learning framework by combination of sparse coding and random forest to identify the informative biomarkers for prediction of clinical scores in SCD and MCI using structural magnetic resonance imaging (MRI). The volumetric features are computed from brain regions and the sub-regions of hippocampus and amygdala in MRIs. Then, sparse coding is applied to identify the relevant features. Finally, the proximity-based random forest is used to combine three sets of volumetric features and establish a regression model for predicting clinical scores. Our method has double feature selections to better explore the relevant features for prediction. Our method is evaluated with the T1-weighted structural MR images from 36 MCI, 112 SCD, 78 Normal Control (NC) subjects. The results demonstrate the effectiveness of proposed method.

Keywords: Subjective cognitive decline · Clinical score prediction · Magnetic resonance image · Random forest

1 Introduction

Mild cognitive impairment (MCI) is a neurological disorder that occurs in older adults involving cognitive impairments. It is often considered as the first clinical precursor of dementia such as Alzheimer's disease (AD) when the individual exhibits lower

© Springer Nature Switzerland AG 2019
I. Rekik et al. (Eds.): PRIME 2019, LNCS 11843, pp. 134–141, 2019.
https://doi.org/10.1007/978-3-030-32281-6_14

performance on standard neuropsychological tests [1]. Recently, a few studies supported that subjective cognitive decline (SCD), which applies to the individuals with self-reported memory complaints, may be the first clinical marker of AD even before MCI [2]. It was shown to have the increased presence of AD biomarkers compared to those without SCD and be associated with a higher risk of progression to AD dementia [3]. Longitudinal studies found that SCD and MCI are associated with a similarly increased risk of AD and predicting rapid cognitive decline [4]. These findings support the idea that SCD may be an early clinical marker of AD that precedes MCI. In order to provide early intervention and delay significant impairment, identification of clinically and cognitively normal individuals who are at risk of AD dementia is very important, especially in the early stage of disease.

Magnetic resonance images (MRI) non-invasively capture the internal body structures, helping us understand the anatomical and functional brain changes related to AD [5]. Some studies have also found that hippocampal atrophy occurs before the onset of AD. A study investigated that SCD individuals have a pattern of hippocampal subfield atrophy similar to that measured in AD pathology when compared to healthy individuals without SCD [6]. The findings indicate the topographically similar changes of hippocampal subfields in SCD individuals as those found in AD. Recently, a study compared SCD with MCI and NC individuals using the volumes and asymmetries of hippocampus, amygdala and temporal horn, and to assess their relationships with cognitive function in elderly population in China [5]. In this study, significant differences ($P < 0.05$) were found in the volumes and asymmetries of both hippocampus and amygdala among the three groups using structural MR images.

The above studies mainly investigated the relationships between the brain atrophy and risk of dementia from SCD, MCI and potential AD through structural MRIs. However, these methods have limitations in exploring the multiple factors on the risk of dementia. With the popularity of machine learning technologies, various methods have been investigated for MR image analysis to find the relevant biomarkers in prediction and analysis of diseases [7]. In addition to the assessment of dementia conditions with sMRI, MMSE and MoCA are often used for initial screening of various types of cognitive impairment and dementia. In fact, NC group has the highest average score in both MMSE and MoCA tests, while these cognitive scores are decreased with the dementia development from SCD, SMCI to AD. Thus, it is necessary to relate the biomarkers of neuroimage to assess and predict MMSE and MoCA scores.

In this work, we investigate the multi-scale brain regions from the ROIs of whole brain to the subregions of hippocampus and amygdala to predict the MMSE and MoCA scores in the early stages of SCD and MCI. We extract three subsets of volumetric features from brain ROIs and the hippocampal and amygdala subregions. The sparse coding is then applied to identify the relevant features for each subset. Finally, the proximity-based random forest is used to combine three sets of volumetric features and establish a regression model for assessment of MMSE and MoCA scores. This study is trying to find the correlation between the volumes of the multi-scale brain regions and the dementia risk to further understand their roles in cognitive impairment and dementia risk. The remainder of this paper is organized as follows. In Sect. 2, we present the materials used in this work and the details of proposed method. Section 3 will present the experimental results and discussion. Finally, we conclude this paper in Sect. 4.

2 Materials and Methods

In this section, we introduce the data set used in this study, followed by the proposed regression method with details. Figure 1 shows the flowchart of our proposed regression framework, which consists of image acquisition and processing, feature extraction and selection, and final score regression.

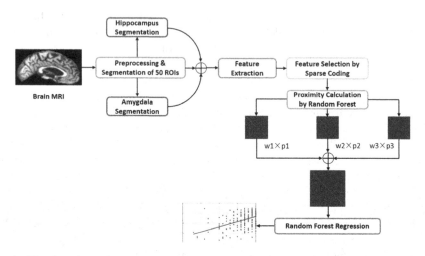

Fig. 1. The flowchart of our proposed regression framework to integrate sparse coding and random forest models for MMSE and MoCA score predictions.

2.1 Materials and Image Processing

The data set in this study are obtained from Shanghai Mental Health Center, China. The participants were recruited from the China Longitudinal Aging Study (CLAS) of Cognitive Impairment (NCT03672448) started in 2011 [8]. This study includes 226 subjects consisting of 36 amnestic MCI, 112 SCD and 78 NC, recruited from a community-based study of individuals aged above 60 in Shanghai, China. Table 1 shows the demographic and clinical information of the studied subjects.

Table 1. Demographic and clinical information of the subjects (Mean ± standard deviation).

Diagnosis	No.	Age	Gender (M/F)	MMSE	MoCA
MCI	36	74.50 ± 7.67	24/12	20.86 ± 3.81	14.47 ± 4.42
SCD	112	70.07 ± 7.58	61/51	27.46 ± 2.27	23.84 ± 4.35
NC	78	67.44 ± 6.41	35/43	28.40 ± 1.56	25.40 ± 3.32

All T1-weighted MR brain images are segmented into 50 regions of interests (ROIs) shown in Table 2 with a fully automated pipeline of FreeSurfer 6.0.0 [9]. The ROI volumes are computed as one subset of features for regression. In addition, the cortex, GM and WM volumes of left and right hemispheres and the volumes of supra tentorial are included in this feature set. There are 57 volumes in this feature set.

Table 2. The segmented 50 ROIs of the whole brain.

Left and right hemisphere		Central region
Cerebellum-white-matter	CA1	Optic-chiasm
Cerebellum-cortex	hippocampal-fissure	CC_Posterior
Thalamus-proper	presubiculum	CC_Mid_Posterior
Caudate	parasubiculum	CC_Central
Putamen	molecular_layer_HP	CC_Mid_Anterior
Pallidum	GC-ML-DG	CC_Anterior
Amygdala	CA3	
Accumbens-area	CA4	
VentralDC	fimbria	
Hippocampal_tail	HATA	
subiculum	Whole_hippocampus	

Furthermore, to investigate the complex structure of hippocampus and amygdala, FreeSurfer is further used to partition these ROIs into 44 and 20 subregions, respectively, as shown in Fig. 2. The volumes are computed from these subregions as two feature sets to predict the cognitive scores.

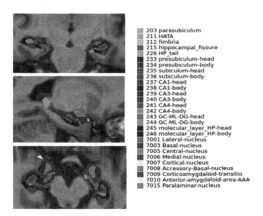

Fig. 2. The segmented subregions of hippocampus and amygdala on one side of brain.

2.2 The Proposed Prediction Method

After segmentation, three subsets of volume features are obtained from the ROIs and the subregions of hippocampal and amygdala to predict the MMSE and MoCA scores. Our proposed method can identify the most relevant features for each subset of features, followed by random forest regression for prediction of clinical scores.

First, sparse coding is used to select the most relevant features for each subset which considers the combination of features over different brain regions to handle the multivariate interactions. Let y denote the clinical scores of training data; A represent the feature matrix of $M \times N$ for M participants; $\vec{\omega} = (\omega_1, \omega_2, \ldots, \omega_N)^T$ is the coefficient vector assigned to the N features. An $L1$-regularized sparsity could be imposed on the coefficients to choose the relevant features for regression. The $L1$-regularized least square problem can be formulated as:

$$\vec{\omega} = \mathrm{argmin}_\omega \left\| y - A\vec{\omega} \right\|_2^2 + \gamma \left\| \vec{\omega} \right\|_1, \quad s.t. \ \vec{\omega}_i \geq 0, \ \forall i \tag{1}$$

where γ is the sparsity regularization parameter which controls the amount of zero coefficients in $\vec{\omega}$. The non-zero elements in $\vec{\omega}$ indicate that the corresponding features are relevant to the regression. The grid search can be used to obtain the optimal sparsity value through cross-validation on the training samples.

Second, random forest [10, 11] is used to compute the proximity measures and make the score regression with the selected features. It can also report the importance of features for each subset. For regression task, decision trees act as regression trees. During the growth of a tree, each node is determined by finding a feature that minimizes the difference between the left and right subset predicting errors. When the predicting error is below a threshold, the node stops splitting as a terminal node. The feature importance can be calculated with the difference between the left and right subset predicting errors. Each weight value is normalized between 0–1. After training, the random forest generates proximity measures showing the probability that two subjects fall into the same leaf node in the regression results of all T trees. Our method has a double feature selection to better explore the relevant features for prediction.

Finally, after 3 individual random forest models are trained to predict the scores with three subsets of features, their proximity matrices are linearly combined into a final proximity matrix as:

$$P = w_1 P_1 + w_2 P_2 + (1 - w_1 - w_2) P_3 \tag{2}$$

where P denotes the final proximity matrix and ω_1, ω_2 are the weights assigned to the corresponding subsets of features. The composite proximity matrix P is input to the random forest model to combine three subsets of features for prediction of scores.

3   Experimental Results

3.1   Datasets and Implementation

The data used in our experiments are from 226 subjects as detailed in Sect. 2.1. In our experiments, the OOB error is converged to stable when $nTree \gtrsim 500$ and the optimal number of trees in the forest $nTree = 1000$. The weighting parameters w_1, w_2 were optimized via grid search in training process to obtain the best performance of random forest regression. The 10-fold cross-validation is used to evaluate the proposed method.

It is repeated 10 times and the final result is obtained by averaging 10 test predictions to reduce the chance of experimental results. To evaluate the prediction performance, we compute the mean squared error (MSE) and the mean absolute error (MAE) between the actual and estimated MMSE and MoCA scores by averaging the results of ten tests. In addition, the Pearson's correlation coefficient (CORR) is used to evaluate the power of regression line in data representation.

3.2   Results on Prediction of Cognitive Scores

The first experiment is to test the effects of different subsets of features on the MMSE and MoCA prediction. We also compare the results by using the *t*-test and sparse coding for feature selection. As for sparse coding, features from 3 subsets are selected separately to get more precise proximity matrix. As for t-test, two groups of data are divided according to the level of scores to select features. The predicting results by using different features and their combinations are listed in Tables 3 and 4, respectively. From the results, we can see the volume features from the subregions of Hippocampus and Amygdala achieve better performances than ROI features. The sparse coding performs better than the *t*-test. Specifically, the proposed combination achieves the highest correlation coefficients of 0.469 and 0.436.

Table 3. The performances comparison for prediction of MMSE scores using different features

Feature	MSE		MAE		CORR	
	t-test	SC	*t*-test	SC	*t*-test	SC
Hippocampus	11.09	10.30	2.48	2.36	0.375	0.408
Amygdala	10.94	10.28	2.46	2.43	0.379	0.429
Whole brain	11.30	11.19	2.46	2.50	0.364	0.375
Combination	10.26	9.68	2.39	2.34	0.434	0.469

Table 4. The performances comparison for prediction of MoCA scores using different features

Feature	MSE		MAE		CORR	
	t-test	SC	*t*-test	SC	*t*-test	SC
Hippocampus	28.48	27.26	4.23	4.09	0.364	0.386
Amygdala	28.31	27.18	4.21	4.16	0.371	0.408
Whole brain	29.20	28.30	4.34	4.30	0.345	0.385
Combination	27.24	25.97	3.99	4.03	0.411	0.436

The second experiment is to test the effects of the weighted combination (WC) of the proximity matrices for fusing three subsets of features on prediction performances. One direct method is to concatenate the selected features from different subsets as the input of regression model. Table 5 shows the prediction performances and the corresponding scatter plots are shown in Fig. 3. We can see that the proposed weighted combination performs better than the concatenating method.

Table 5. Performance comparison for prediction of clinical scores with different combinations

	Feature	MSE	MAE	CORR	Predicted	True score
MMSE	Concatenate	9.68	2.34	0.469	26.70 ± 1.46	26.72 ± 3.52
	WC	9.55	2.31	0.500	26.70 ± 1.54	
MOCA	Concatenate	25.97	4.03	0.436	22.72 ± 2.28	22.81 ± 5.67
	WC	25.61	4.01	0.450	22.77 ± 2.50	

3.3   Biomarkers Relevant to the Predictions of Cognitive Scores

In this section, we investigate the relevant biomarkers for disease interpretation. We computed the number of times that the features were selected out of 10 folds and denoted as frequency. The features with frequency higher than 8 were selected as the relevant biomarkers for each partition. Our study found that hippocampus atrophy in the right hemisphere has a higher weight than the left on the scores while the amygdala is just the opposite. The hippocampal fimbria shows the highest weight among all ROIs, with right fimbria showing higher weight than the left. The results indicate that the commonly selected top regions are consistent to the AD pathology studies [5, 6, 12].

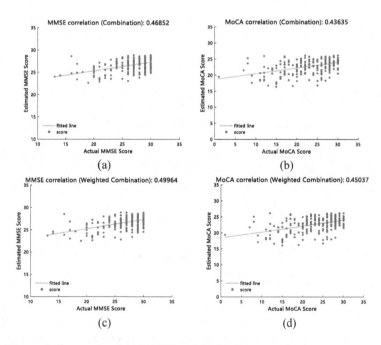

Fig. 3. The prediction results of (a) MMSE and (b) MoCA scores by feature concatenating, as well as the prediction results of (c) MMSE and (d) MoCA scores by weighted combination.

4   Conclusion

In this paper, we have proposed a combined regression framework based on sparse coding and random forest for prediction of MMSE and MoCA scores. It enables MRI diagnostic analysis of the SCD group, which is rarely involved in current research. Three sets of volumetric features are extracted from the ROIs of whole brain and the subregions of hippocampus and amygdala. Sparse coding is applied to select the relevant features to clinical score estimation. As for brain ROIs, the paper subdivided the subregions on the basis of the hippocampus and the amygdala. By comparison with the whole brain, it is proved that the amygdala is more closely associated with clinical scores, followed by hippocampus. These results are also consistent with relative clinical experiments, achieving computer-aided diagnosis and prediction of AD process through the calculation and analysis of brain MRI.

References

1. Silveira, M., Marques, J.: Boosting Alzheimer disease diagnosis using PET images. In: 2010 20th International Conference on Pattern Recognition, pp. 2556–2559. IEEE, (2010)
2. Lin, Y., Shan, P.-Y., Jiang, W.-J., Sheng, C., Ma, L.: Subjective cognitive decline: preclinical manifestation of Alzheimer's disease. Neurol. Sci. **40**, 41–49 (2019)
3. Tales, A., Jessen, F., Butler, C., Wilcock, G., Phillips, J., Bayer, T.: Subjective cognitive decline. J. Alzheimers Dis. **48**, S1–S3 (2015)
4. Kirkova, V., Traykov, L.: Predictors of cognitive decline and dementia in individuals with subjective cognitive impairment: a longitudinal study. J. Neurol. S42 (2013). Springer, Heidelberg Tiergartenstrasse 17, D-69121 Heidelberg, Germany (2013)
5. Yue, L., et al.: Asymmetry of hippocampus and amygdala defect in subjective cognitive decline among the community dwelling Chinese. Front. Psychiatry **9** (2018)
6. Perrotin, A., et al.: Hippocampal subfield volumetry and 3D surface mapping in subjective cognitive decline. J. Alzheimers Dis. **48**, S141–S150 (2015)
7. Liu, M., Cheng, D., Wang, K., Wang, Y., Alzheimer's Disease Neuroimaging Initiative: Multi-modality cascaded convolutional neural networks for Alzheimer's disease diagnosis. Neuroinformatics **16**, 1–14 (2018)
8. Xiao, S., et al.: Methodology of China's national study on the evaluation, early recognition, and treatment of psychological problems in the elderly: the China Longitudinal Aging Study (CLAS). Shanghai Archives of Psychiatry **25**, 91 (2013)
9. Fischl, B.: FreeSurfer. Neuroimage **62**, 774–781 (2012)
10. Breiman, L.: Random forests. Mach. Learn. **45**, 5–32 (2001)
11. Svetnik, V., Liaw, A., Tong, C., Culberson, J.C., Sheridan, R.P., Feuston, B.P.: Random forest: a classification and regression tool for compound classification and QSAR modeling. J. Chem. Inf. Comput. Sci. **43**, 1947–1958 (2003)
12. Evans, T.E., et al.: Subregional volumes of the hippocampus in relation to cognitive function and risk of dementia. Neuroimage **178**, 129–135 (2018)

Diagnosis of Parkinson's Disease in Genetic Cohort Patients via Stage-Wise Hierarchical Deep Polynomial Ensemble Learning

Haijun Lei[1], Hancong Li[1], Ahmed Elazab[2], Xuegang Song[2],
Zhongwei Huang[1], and Baiying Lei[2(✉)]

[1] Key Laboratory of Service Computing and Applications,
Guangdong Province Key Laboratory of Popular High Performance Computers,
College of Computer Science and Software Engineering, Shenzhen University,
Shenzhen 518060, China
[2] National-Regional Key Technology Engineering Laboratory for Medical
Ultrasound, Guangdong Key Laboratory for Biomedical Measurements
and Ultrasound Imaging, School of Biomedical Engineering,
Health Science Center, Shenzhen University, Shenzhen 518060, China
leiby@szu.edu.cn

Abstract. As a neurodegenerative disease, Parkinson's disease (PD) has gradually become common in the elderly. Effective disease diagnosis has become increasingly important, especially in the patients with mutation of PD related gene. Due to the slight changes in the brain, it is very difficult to diagnose PD by neuroimaging techniques. In order to be more effective in assisting diagnosis, we further improve the deep polynomial network (DPN) as the hierarchical stacked DPN (HSDPN) and propose a stage-wise hierarchical deep polynomial ensemble learning (SHDPEL) framework for encoding multiple features to obtain high-level feature representations of different neuroimaging segmentation in PD diagnosis. Specifically, we train different segmentation features separately in the first stage. In next stage, different combinations of feature pairs will be used to learn the correlative information between different segmentations. We further integrate all branches by using a voting ensemble strategy for the classification. A series of experiments are performed on all the neuroimaging data to demonstrate the effectiveness of this method on the publicly available Parkinson's Progression Marker Initiative (PPMI) dataset. The experimental results show that the method can achieve remarkable results and is superior to related methods.

Keywords: Parkinson's disease · Genetic cohort · Hierarchical stacked deep polynomial network · Stage-wise learning · Ensemble learning

This work was supported partly by the Integration Project of Production Teaching and Research by Guangdong Province and Ministry of Education (No. 2012B091100495), Shenzhen Key Basic Research Project (No. JCYJ20170302153337765) and Guangdong Pre-national Project (No. 2014 GKXM054).

© Springer Nature Switzerland AG 2019
I. Rekik et al. (Eds.): PRIME 2019, LNCS 11843, pp. 142–150, 2019.
https://doi.org/10.1007/978-3-030-32281-6_15

1 Introduction

At present, the diagnosis of Parkinson's disease (PD) depends mainly on clinical symptoms [1]. It requires a great deal of knowledge and expertise from the clinician. In addition, the subjective evaluation may vary from clinician to another [1]. In order to improve PD diagnostic performance, the effective computer-aided diagnosis (CAD) systems can play an important role. As an important auxiliary tool in clinical diagnosis, neuroimaging can provide a data foundation for CAD [2]. Researchers have used neuroimaging (e.g., magnetic resonance imaging (MRI)) for automated PD diagnosis as it can reveal structural abnormalities in the brain. Furthermore, genes play an increasingly important role in the diagnosis of PD [3]. Although unaffected individuals with genetic mutations have not developed PD-related clinical symptoms or any reduction in dopamine, they also have the potential to develop into PD [4]. Effectively distinguishing PD patients from other categories of subjects will help to monitor their health statuses. Therefore, the physical condition of these unaffected individuals with genetic mutations also needs to be highly concerned.

For the diagnosis of PD, deep learning (DL) can be a potential tool since it has been successfully used in various fields of medical image analysis and achieve quite promising performance [5, 6]. However, DL algorithm using limited annotated medical imaging data may be undesirable due to its huge amount of learning parameters and the prone to over-fitting. Particularly, the data insufficiency is even more challenging as it is very difficult to acquire data with genetic mutations. To address this, researchers have been working on data transformations to increase the feature space for performance boosting even with the small amount of data [7]. Currently, most study of PD diagnosis based on regions-of-interest (ROI) are feature selection methods via machine learning [8–10]. However, multiple parameters need to be adjusted to achieve better performance in feature selection methods. The establishment of the model using only the selected brain regions largely ignores the information provided by the remaining brain regions.

In current study, since the dataset with genetic mutations is much smaller than the dataset without genetic mutations, we decide to develop the deep polynomial network (DPN) method to complete the diagnosis of PD, which can effectively utilize the small dataset [11]. Experimental results of some commonly used large-scale image datasets show that DPN method has competitive or even better performance than the other basic DL algorithm [11], such as deep belief network (DBN) [12], stacked auto-encoder (SAE) [13]. In this respect, we improve the DPN as hierarchical stacked DPN to learn better feature representations from small datasets. In addition, we proposed the stage-wise hierarchical deep polynomials ensemble learning method to discover more discriminative information between different features in limited data. Experiments on the public Parkinson's Progression Marker Initiative (PPMI) [14] dataset shows the promising performance achieved for various classification tasks.

2   Methodology

2.1   Data Preprocessing

To preprocess the MRI data, we first apply the anterior commissure-posterior commissure reorientation in all MRI data using the center of a mass algorithm. Then, we correct the head movement and geometric distortion by using the statistical parametric mapping toolbox (SPM8). In the next step, we implement a graph-cut method [15] for skull-stripping. We also employ the international consortium for brain mapping template to register the data, which serves as a reference for providing coordinates and divides the specific and associated regions into gray matter (GM), white matter (WM) and cerebrospinal fluid (CSF). Afterward, the automated anatomical labeling (AAL) atlas, which is a predefined 116 ROIs template of the human brain, is used for the GM, WM and CSF registration, respectively. Specifically, a total of 116 ROIs are partitioned from GM, WM, and CSF and spatially normalized by the AAL atlas with high-resolution 3D brain atlas. At last, we extract the feature from each ROI as the input data of our method.

2.2   Deep Polynomial Network

For DPN algorithm, it learns the polynomial predictor through a deep network structure, which provides a good approximation basis for the values obtained by the polynomial on the training samples. The basic DPN block is shown in Fig. 1(a). Let $\{(x_i, y_i)\}_{i=1}^{n}$ denote the training samples, we use $(w_1, w_2, \ldots, w_n)$ and $(p_1, p_2, \ldots, p_n)$ respectively to represent the coefficient vector and n polynomials in the DPN network.

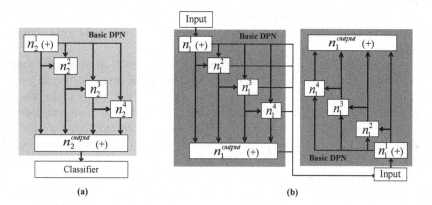

Fig. 1. (a) Basic DPN block of $4°$, (b) Hierarchical Stacked DPN with two basic DPN block: the hierarchical output of first DPN block is connected to the input of the next basic DPN block.

Then, we obtain $\sum_{i=1}^{n} w_i p_i(x_j) = y_j$ for any output values $(y_1, y_2 \ldots, y_n)$ according to the Lemma in [11]. The first-degree polynomials are as follow:

$$\{(<w,[1x_1]>,\ldots,<w,[1x_n]>) : w \in \mathbb{R}^{d+1}\}. \tag{1}$$

Since we have the degree-1 polynomial, we use a singular value decomposition method to search a series of $(d+1)$-dimension vectors $(w_1, w_2, \ldots, w_{d+1})$. We finally map $[1X]$ to the constructed basis via linear transformation. Let $L^1 \in \mathbb{R}^{m \times (d+1)} = <W_j, [1X]>$ be its columns and linearly independent, the construction of the first layer L^1 and matrix $[L\tilde{L}^2]$ is:

$$\tilde{L}^2 = \left[\left(L_1^1 \circ L_1^1 \right) \ldots \left(L_1^1 \circ L_{|L_1|}^1 \right) \ldots \left(L_{|L_1|}^1 \circ L_1^1 \right) \ldots \left(L_{|L_1|}^1 \circ L_{|L_1|}^1 \right) \right], \tag{2}$$

where $F_1^1 \circ F_1^1$ is the Hadamard product [16]. The matrix $[L\tilde{L}^2]$ is obtained by the second degree polynomials and its columns are linearly independent while the L^2 designates the 2nd-layer network.

So far, L is redefined as the augmented matrix $[LL^2]$. We can extend to construct $3^{rd}, \ldots, z^{th}$ layer by repeating the same process and constructing the 2nd layer of the network. Any other degree-z can be represented as the degree-1 polynomials and degree-$(z-1)$ polynomials. Hence, we reformulate the $\tilde{F}^z$ matrix as:

$$\tilde{L}^z = \left[\left(L_1^{z-1} \circ L_1^1 \right) \ldots \left(L_1^{z-1} \circ L_{|L_1|}^1 \right) \ldots \left(L_{|L^{z-1}|}^1 \circ L_1^1 \right) \ldots \left(L_{|L^{z-1}|}^1 \circ L_{|L_1|}^1 \right) \right], \tag{3}$$

where $[LL^2]$ is from a basis $[L\tilde{L}^z]$. To maintain numerical stability, $\tilde{L}^z$ is converted to L^z via:

$$L_r^z := W_{i(r),j(r)} L_{i(r)}^{z-1} \circ L_{j(j)}^1, r = 1, \ldots, |L^z|. \tag{4}$$

After building up the main network structure, we can now train a simple linear predictor w and minimize some convex loss function $w \to l(L_w, y)$. This can be done using any convex optimization procedure. Afterward, we output the feature from the last layer of the basic DPN network. To avoid overfitting issues, the network width is set equal to the number of columns in L^t, with explicit constraints on each iteration as suggested in [16]. To calculate the first layer in the DPN, the linear transformation transforms the enhanced data matrix $[1X]$ to the top k leading singular vectors by using the principal component analysis. In the next layer of construction networks, a standard orthogonal least squares algorithm is employed to iteratively select the most relevant residuals, which is most relevant to the actual label of data.

2.3 Hierarchical Stacked Deep Polynomials Network

In order to learn deeper semantic features in limited data, we proposed the hierarchical stacked DPN (HSDPN) algorithm based on SDPN [17] as the key module in our framework, where multiple basic DPNs with hierarchical feature output from each layer in the DPN block. Figure 1(b) shows an example of HSDPN model with two 4-degrees basic DPNs. It can be stacked on each other to construct a deep structure. The

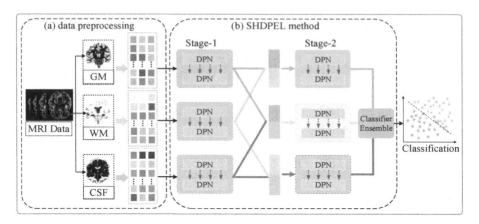

Fig. 2. Proposed framework of stage-wise hierarchical deep polynomials ensemble learning.

hierarchical learned features from different layers are then fed to next DPN block to further improve the feature representation. Afterward, the output of the last DPN block will be used in the classifier.

As we mentioned in the previous Section, each basic DPN is trained in a supervised network, which sticks to the block-by-block manner without the need for back-propagation. It is also worth noting here that, the number of network layers and nodes of each DPN block can be set separately, which makes HSDPN quite flexible for various applications. In addition, advanced features can be learned by the SDPN. Therefore, the HSDPN should be superior to the basic DPN in the feature representation.

2.4   Stage-Wise Hierarchical Deep Polynomials Ensemble Learning

Since we extract three types of features from MRI data to fuse and learn different kind of feature representations, we can put these features into the DPN through a simple feature fusion. However, this simple connection strategy neglects to some extent the diversity among multiple features, and it does not well explore complementary nature and represents the high nonlinear correlation among multiple features. Furthermore, simple connection might ignore the related information among different features. In order to attain a better performance, we propose stage-wise hierarchical deep poly-nomial ensemble learning (SHDPEL) method to integrate the multi-level features. In addition, our method performs a stage-wise feature learning strategy on different fea-tures of MRI data. The overall structure of SHDPEL method which contains HSDPN modules with several basic DPN blocks is illustrated in Fig. 2. In the first stage, feature learning is performed on single type of features and post-learning features are obtained. The discriminative features of each MRI data can be used as input to its corresponding HSDPN module to learn the advanced feature representation, respectively. Afterward, we separately combine the processed features as different feature pairs (GM-WM, GM-CSF, and WM-CSF) for the second stage training. The HSDPN module in the second stage correlates feature information for different combinations of feature pairs. The aim of this stage is to fuse the complementary information from different feature to further improve the performance of the classification framework.

Furthermore, we employ the support vector machine (SVM) classifier with the sigmoid kernel to accomplish the classification of different combination in the second stage. Since each feature pairs contributes uniquely to the final prediction and the majority voting is able to enhance the performance by exploring the feature complementarity, we compute the weighted summation (voting) of the prediction results in different combinations as the final classification result. Therefore, the features learned by SHDPEL method can be more discriminative and robust.

3 Experiments and Results

3.1 Dataset and Experimental Setting

In our work, all the used subjects are from the PPMI dataset. There are 170 subjects including 38 genetic cohort unaffected individuals (GenUn), 42 genetic cohort PD (GenPD), 56 scans without evidence of dopaminergic degeneration (SWEDD), and 34 Prodromal subjects. Note that, all the subjects in GenUn are not PD patient. Also, all MRI data used in the current study are the T1 weighted MRI images.

In this paper, we only use the baseline data for the evaluation of SHDPEL method. Meanwhile, we perform four binary classification tasks, namely GenUn vs. GenPD, GenUn vs. SWEDD, GenUn vs. Prodromal, and SWEDD vs. Prodromal. The first three groups of classification task are for the normal group and the group with clinical symptoms and pathological changes. Prodromal and SWEDD are special groups between normal people and PD, so it is necessary to effectively distinguish between them.

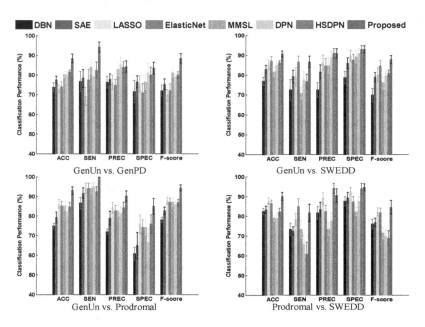

Fig. 3. Classification performance of all method in 4 classification tasks.

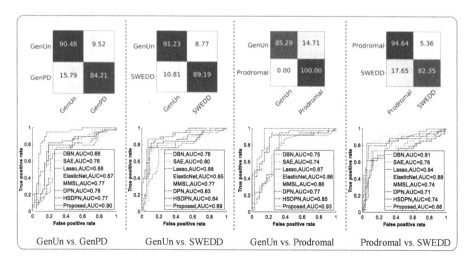

Fig. 4. Confusion matrix of our proposed method and ROC curves of all method in 4 classification tasks.

In every single task, we use the 10-fold cross-validation strategy to compare with all competing methods. Specifically, at first, we randomly divide the entire dataset into 10 subsets. Then, we choose 1 subset to be the test set and use the remaining 9 subsets for training. We repeat the entire process 10 times to avoid possible bias in cross-validation during dataset partitioning. In order to estimate the classification performance of our proposed method, we use quantitative measurements including accuracy (ACC), sensitivity (SEN), specificity (SPEC), precision (PREC), and F-score (F1).

3.2   Classification Performance

In this paper, the SHDPEL method is compared with both the deep learning methods (DBN and SAE) and the feature selection methods via machine learning (Least absolute shrinkage and selection operator (LASSO) [10], ElasticNet [9] and Multi-modal sparse learning (MMSL) [8]). Specifically, MMSL is an improved version of multi-modal multi-task learning (M3T) [18]. We use a simple linear combination to fuse different features from GM, WM, and CSF, which is the input for all the comparisons. For fair evaluation, all methods use the sigmoid kernel SVM as the classifier. In addition, we use DPN and HSDPN which are the same network structure compare to the DPN in our proposed method to further verify the performance.

In all experiments, we perform the SHDPEL method on four sub-classification tasks and compare the results with the methods mentioned in the previous Section. The experimental results of all method can be viewed in Fig. 3. In Fig. 4, we show the confusion matrix results achieved by the proposed method for all binary classification tasks. Note that we report the final confusion matrices by averaging the result of the 10-fold cross validation. In addition, Fig. 4 shows various receiver operating characteristic (ROC) curves of all the results. These results show good performance achieved by our proposed method, which is indicated by the red dotted line. In the first three groups of experiment, SHDPEL method has clear advantage in classification performance. In addition, our methods with DPN modules are superior in performance to both deep learning and feature selection methods. We also achieve the best results in all DPN related methods in first 3 groups (e.g., the accuracies of 88.49%, 90.67%, and 92.98%).

Figure 3 also shows the classification of Prodromal vs. SWEDD. Both categories are subject with PD related symptoms between the PD and the normal subjects. The clinical signs of PD are examined in the SWEDD subjects, but no reduction in dopamine is detected, which is just the opposite to the Prodromal. Effectively distinguishing these two intermediate states will have an auxiliary role in the early diagnosis and treatment of PD. As can be noticed, our proposed method (with an accuracy of 90.11%) achieves the best performance, and it demonstrates that our proposed method has great potentiality in feature learning and representation. Furthermore, all the experimental results confirm the applicability of our method to use in the small sample dataset. Even if the sample in the genomic group is much smaller, we can effectively train our network and achieve better performance than competing methods.

4   Conclusion

In this paper, we propose a stage-wise feature learning based classification framework to generate advanced feature representations from neuroimaging data for diagnosis in PD. The experimental results show that the SHDPEL method has excellent performance in the small-scale Parkinson's dataset, indicating that our HSDPN block is very suitable for learning multi-layer feature representation of small medical datasets. The stage-wise learning strategy is able to find out the correlation among different features and also effectively learn more high-level information in neuroimaging data. In the future, we will consider different multimodalities as our training data and further improve our network structure to increase diagnosis performance.

References

1. Tysnes, O.B., Storstein, A.: Epidemiology of Parkinson's disease. J. Neural Transm. **124**, 1–5 (2017)
2. Nichols, T.E., et al.: Best practices in data analysis and sharing in neuroimaging using MRI. Nature Neurosci. **20**, 299 (2017)
3. Hernandez, D.G., Reed, X., Singleton, A.B.: Genetics in Parkinson disease: mendelian versus non-mendelian inheritance. J. Neurochem. **139**, 59–74 (2016)
4. Poewe, W., et al.: Parkinson disease. Nat. Rev. Dis. Prim. **3**, 17013 (2017)
5. Liao, S., Gao, Y., Oto, A., Shen, D.: Representation learning: a unified deep learning framework for automatic prostate MR segmentation. Med. Image Comput. Comput. Assist. Interv. **16**, 254–261 (2013)
6. Hoo-Chang, S., Orton, M.R., Collins, D.J., Doran, S.J., Leach, M.O.: Stacked autoencoders for unsupervised feature learning and multiple organ detection in a pilot study using 4D patient data. IEEE Trans. Pattern Anal. Mach. Intell. **35**, 1930–1943 (2013)

7. Li, D.C., Liu, C.W., Hu, S.C.: A fuzzy-based data transformation for feature extraction to increase classification performance with small medical data sets. Artif. Intell. Med. **52**, 45–52 (2011)

8. Lei, H., et al.: Joint detection and clinical score prediction in Parkinson's disease via multimodal sparse learning. Expert Syst. Appl. **80**, 284–296 (2017)

9. Zou, H., Hastie, T.: Regularization and variable selection via the elastic net. J. R. Stat. Soc.: Ser. B (Stat. Methodol.) **67**, 301–320 (2005)

10. Tibshirani, R.: Regression shrinkage and selection via the lasso: a retrospective. J. R. Stat. Soc.: Ser. B (Stat. Methodol.) **73**, 273–282 (2011)

11. Livni, R., Shalevshwartz, S., Shamir, O.: An algorithm for training polynomial networks. Comput. Sci. **26**, 4748–4750 (2013)

12. Hinton, G.E., Osindero, S., Teh, Y.-W.: A fast learning algorithm for deep belief nets. Neural Comput. **18**, 1527–1554 (2006)

13. Suk, H.I., Lee, S.W., Shen, D.: Latent feature representation with stacked auto-encoder for AD/MCI diagnosis. Brain Struct. Funct. **220**, 841–859 (2015)

14. Marek, K., et al.: The parkinson progression marker initiative (PPMI). Prog. Neurobiol. **95**, 629–635 (2011)

15. Sadananthan, S.A., Zheng, W., Chee, M.W.L., Zagorodnov, V.: Skull stripping using graph cuts. Neuroimage **49**, 225–239 (2010)

16. Fill, J.A., Flajolet, P., Kapur, N.: Singularity analysis, Hadamard products, and tree recurrences. J. Comput. Appl. Math. **174**, 271–313 (2005)

17. Shi, J., Zhou, S., Liu, X., Zhang, Q., Lu, M., Wang, T.: Stacked deep polynomial network based representation learning for tumor classification with small ultrasound image dataset. Neurocomputing **194**, 87–94 (2016)

18. Zhang, D., Shen, D., Alzheimer's disease neuroimaging initiative: multi-modal multi-task learning for joint prediction of multiple regression and classification variables in Alzheimer's disease. NeuroImage **59**, 895–907 (2012)

Automatic Detection of Bowel Disease with Residual Networks

Robert Holland[1], Uday Patel[2], Phillip Lung[2], Elisa Chotzoglou[1],
and Bernhard Kainz[1(✉)]

[1] Department of Computing, Imperial College London, BioMedIA, London, UK
{robert.holland15,e.chotzoglou16,b.kainz}@imperial.ac.uk
[2] St. Mark' Radiology, London North West University Healthcare NHS Trust,
London, UK
{udaypatel2,phill15plung}@nhs.net

Abstract. Crohn's disease, one of two inflammatory bowel diseases (IBD), affects 200,000 people in the UK alone, or roughly one in every 500. We explore the feasibility of deep learning algorithms for identification of terminal ileal Crohn's disease in Magnetic Resonance Enterography images on a small dataset. We show that they provide comparable performance to the current clinical standard, the MaRIA score, while requiring only a fraction of the preparation and inference time. Moreover, bowels are subject to high variation between individuals due to the complex and free-moving anatomy. Thus we also explore the effect of difficulty of the classification at hand on performance. Finally, we employ soft attention mechanisms to amplify salient local features and add interpretability.

1 Introduction

1.1 Motivation

Most people suffering from Crohn's disease are younger than 35 and the cost of their treatment exceeds £500 million in the UK alone. Symptoms include inflammation of tissue anywhere along the gastrointestinal tract. However, it is most commonly found in the terminal ileum (where the small and large intestine meet). While there is no cure, early detection can vastly improve quality of life.

A successful algorithm would assist radiologists in more accurate diagnosis and follow-up of Crohn's disease. This would be of particular benefit to radiologists with limited experience of Crohn's disease imaging or who encounter patients with Crohn's disease uncommonly. Such an algorithm could also be used to triage patients so that severe cases can be reviewed more immediately, or to perform a secondary review to the radiologist and flag potentially missed cases.

© Springer Nature Switzerland AG 2019
I. Rekik et al. (Eds.): PRIME 2019, LNCS 11843, pp. 151–159, 2019.
https://doi.org/10.1007/978-3-030-32281-6_16

1.2   Study Outline and Contributions

Performance of classification tasks on the bowels is degraded by the intrinsic complexity and noise of the anatomy. While Crohn's disease can inflame the entire gastrointestinal (GI) tract, radiologists typically study the terminal ileum when making a diagnosis [1]. The first question we consider is the extent to which is it possible to classify IBD Crohn's disease from an MRI volume using vanilla deep learning methods. To establish this baseline, we first localise to the ROI using the patient-specific coordinates of the terminal ileum provided by a radiologist. We demonstrate that this semi-automatic technique performs comparably to the current standard for evaluating Crohn's with MRI, the MaRIA score [8], while requiring only a fraction of the preprocessing. We also explore how the difficulty and inflammation severity of a sample affects classification performance.

The assumption will then be dropped, such that we are forced to work only with population-specific knowledge, resulting in weaker localisation. Precision and recall degrade as the now fully-automatic algorithm encounters a worse signal-to-noise ratio (SNR). Finally, we show that in the absence of overfitting soft attention mechanisms [9] improve performance through amplification of salient local features.

2   Related Work

Currently there are no deep learning methods deployed in the clinic to assist diagnosis of Crohn's disease. Diagnosis is determined entirely by radiologists and clinical professionals who employ various *in vivo* and imaging techniques. Thus, our classification performance will be compared with the clinical standard, the MaRIA score. For their similarity in physical domain, we then review similar applications of deep learning to the abdomen.

2.1   Clinical Standards for Evaluating IBD

The first methods to standardise diagnosis of IBDs were endoscopic scoring systems, such as the Crohn's Disease Endoscopic Index of Severity (CDEIS). However, these incur practical issues; regular endoscopic examinations have several drawbacks related to '*invasiveness, procedure-related discomfort, risk of bowel perforation and relatively poor patient acceptance*' [8]. In fact, a meta-analysis of prospective studies has shown both MRE and CT to have a sensitivity and specificity of greater than 90% in diagnosing IBDs. To evaluate the MRI, radiologists visually examine the bowels slice by slice and look for high level features. Signs indicative of IBD include increase in T2 signal and thickness of the bowel walls. Rimola et al. [8] developed a scoring system, the MaRIA score, by first extracting these standardised imaging features through manual annotation by a radiologist and then fitting them in a regression model. MaRIA score was found to have a strong correlation with CDEIS. For the detection of disease activity it scored **0.81** for sensitivity and **0.89** for specificity.

Challenges in computing the MaRIA score include differentiation of diseased segments from those that are collapsed, variability of disease presentation and image degradation caused by motion [2]. Additionally, the aforementioned metrics used in the MaRIA score must be calculated by a radiologist in the terminal ileum, the transverse, ascending, descending and sigmoid colon and the rectum, which is a timely and costly procedure.

2.2   Machine Learning for the Automated Detection of IBD

Machine learning can automatically extract local features in the presence of noise, and combine them to make more complex decisions. Thus, it promises to automate the collection of low level features and, as we determine in this work, the diagnosis. Some attempt has been made to automate the collection of features specifically for calculation of the MaRIA score; in 2013 Schüffler et al. [7] used random forests to segment diseased bowels. However, this technique first requires a radiologist to indicate the section of diseased bowel to evaluate. Moreover at the time of the study it required one hour per patient. As far as we can see, there are no studies that use deep learning to directly diagnose IBD from imaging data. Moreover, there are comparatively few medical imaging challenges that focus on the abdomen (notably KiTS19 and CHAOS19) compared to other domains, and as far as we can see, none that regard IBD.

Typically, the medical imaging community has been more focused on tasks such as tumour, lesion and anatomical segmentation. This is evidenced in 'A Survey of Deep Learning in Medical Image Analysis' [5], detailing that *'Most papers on the abdomen aimed to localize and segment organs, mainly the liver, kidneys, bladder, and pancreas'*. A more recent review paper, 'An overview of deep learning in medical imaging focusing on MRI' [6], describes continued progress in segmentation, registration and image synthesis, but regarding diagnosis and prediction it advises to consult the list from the previous review [5] indicating that the main focus still lies in segmentation. Indeed, newer studies on the task of prediction and diagnosis concentrate on the brain, kidney, prostate and spine, but do so via segmentation rather than direction prediction.

Thus, it may be the case that the optimal method for diagnosing Crohn's IBD operates by first segmenting the terminal ileum. Abdominal segmentation has been attempted, though not including the terminal ileum [3]; dice scores were high for larger anatomy (e.g. liver at 95.3 ± 0.7) but significantly reduced for smaller anatomy similar in function to the terminal ileum (e.g. duodenum at 65.5 ± 8.9). Furthermore, they go on to describe the limitations of CNNs for inference in the bowels, commenting that *'It is very challenging for the CNN to learn stable representative features for the digestive organs because the appearances, shapes, and sizes of these organs are highly unstable from day to day depending on different food intake and digestion process'* [3].

To summarise, there are no studies making direct diagnosis of IBD using deep learning on images. Furthermore, there are also no learning algorithms since the random forests [7] diagnosing IBD from MR volumes. Segmentation is typically preferred to direct diagnosis due to the increased dimensionality of the annotations. As such, we compare our baseline performance to the reported binary classification performance of the MaRIA score in classification of Crohn's disease.

3 Data

MRI data has been acquired on a Philips Achieva 1.5 T MR System with acquisition parameters as outlined in Table 1. Use of de-identified data has been consented by the local ethics committees.

Table 1. MRI acquisition parameters. Number of signal averages (NSA); Turbo spin echo (TSE)

Planes	Sequence	FOV [mm]	TR/ TE	Slice [mm]	Matrix	NSA	Time [s]
Axial	e-THRIVE (T1 FFE / TFE)	375	5.9/3.4	3	212×160	1	20.7×2
Coronal	Single shot TSE (T2 TSE)	375	554/120	3	300×213	1	21.1
Axial	Single shot TSE (T2 TSE)	375	587/120	3.5	304×255	1	22.3×2

The Crohn's MRI dataset is divided into healthy, mild, moderate and severe (with fistulation) terminal ileal inflammation. These represent severity levels 0, 1, 2 and 3 respectively which were originally calculated using the MaRIA score. As there are no terminal ileal ground-truth segmentations available, the only other annotation is the centroid coordinates of the terminal ileum.

Individuals are ranked by classification *difficulty*; an ordering determined by the radiologists who annotated the data. While we cannot formally describe how an MRI volume of a patient might be *difficult* to annotate, we can theorise that it means the symptoms of Crohn's disease are hard to spot or are borderline. These difficulties may correspond to those discussed in computing the MaRIA score discussed in Sect. 2.1. Indeed, we see that as the severity of inflammation decreases the difficulty increases in Table 2 (average difficulty is 35.0).

Table 2. Distribution of inflammation and suggested classification difficulty

Inflammation class	Frequency	Average difficulty
Healthy	100	N/A
Mild	34	39.1
Moderate	29	35.3
Severe	7	19.1

Formally, let $\{\mathbf{i}_j\}_{j=1}^N, \{\mathbf{d}_j\}_{j=1}^N$ such that $\forall j\ \mathbf{i}_j, \mathbf{d}_j \in \mathbb{R}^3$ be the set of physical locations of the terminal ilea and the dimensions of the jth patient respectively. Then let the proportional ileal location be $\mathbf{p}_j = \frac{\mathbf{i}_j}{\mathbf{d}_j}$ and suppose that

$$\forall j\ \mathbf{p}_j \sim \mathcal{N}(\mu, \Sigma)$$

The distribution of $\{\mathbf{p}_j\}_{j=1}^N$ is shown in Fig. 1. Given that $\hat{\mu} < 0$ and $\hat{\Sigma}$ is small we observe that the terminal ileum is usually confined to one octant of the volume. From this distribution we can define a bounding box that we expect to contain all ilea. We make use of this assumption in preprocessing (see Sect. 3.1). We also observe from $\hat{\Sigma}$ that most variation is in the axial direction. This is expected as the method by which we determine the patient's size is most uncertain in this direction (patient dimensions were determined by region growing).

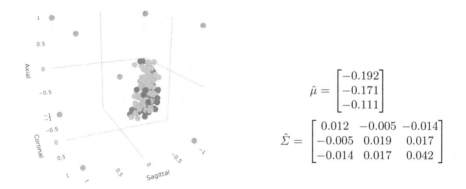

$$\hat{\mu} = \begin{bmatrix} -0.192 \\ -0.171 \\ -0.111 \end{bmatrix}$$

$$\hat{\Sigma} = \begin{bmatrix} 0.012 & -0.005 & -0.014 \\ -0.005 & 0.019 & 0.017 \\ -0.014 & 0.017 & 0.042 \end{bmatrix}$$

Fig. 1. Terminal ileal population distribution (normalised to $[-1, 1]$)

3.1 Application Variants

We can localise to the ROI using the coordinates of the terminal ileum by extracting a small surrounding window, resulting in the *Localised* dataset. However, in the fully-automatic variant, we are forced to extract a larger region using the estimated distribution shown in Fig. 1 resulting in the *Generic* dataset. The effect of localisation strength on performance is detailed in Sect. 5. Localisation is crucial for mitigating overfitting, but also for permitting larger batch sizes, since the *Generic* and *Localised* techniques result in 95.8% and 99.4% volume reductions respectively.

4 Method

We are interested in the binary classification power of vanilla deep learning frameworks. As such, due to its efficient use of parameters, we chose ResNet [4]

- this affords us larger batch sizes, which are restricted by the dimensionality of the scans. Our custom Resnet uses exclusively 3^3 filters and ReLU activation. Refer to our network specification in Table 3. Each set of residual blocks, $\mathbf{d}_j$, begins with a downsampling layer via strided convolution. The residual blocks are followed by a classification module, comprising a global average pooling layer which allows us to feed inputs of variable size to the network. It also reduces the number of learnable parameters in the model as it is followed by a dropout fully connected layer resulting in two output neurons, as in binary classification.

We also add soft attention layers as described in Attention-gated Sononet by Schlemper et al. [9]. These act as a gate for signal by learning the compatability between pixel-wise features at a large scale and more global, discriminative features taken before the final soft-max layer. This is then normalised to form the attention map (see Fig. 2 for examples) and the dot product is taken with the pixel-wise features to produce attended features. These too pass through a classification module and their prediction is weighted against that of the original network's. To extend our custom Resnet we add an attention layer before the final downsampling layer. This multi-scale technique assists the network in identifying local, salient features such as the terminal ileum and is shown to improve performance in the absence of overfitting.

Table 3. Our ResNet configuration for input volume of size $31 \times 87 \times 87$

Layer	Channels	Blocks	Resultant feature map
$\mathbf{d}_1$	64	3	$16 \times 44 \times 44$
$\mathbf{d}_2$	128	3	$8 \times 22 \times 22$
$\mathbf{d}_3$	256	3	$4 \times 11 \times 11$
Global average pooling			256
Dense layer			2

4.1 Training and Evaluation

Loss is computed as cross entropy between the logits and the ground truth labels. We use Adam with $\beta_1 = 0.9, \beta_2 = 0.99$ for the first and second order moment coefficients respectively, and a learning rate of $5 \cdot 10^{-6}$. Due to the reduction in volume (see Sect. 3.1) we can use batch sizes of 64, a significant portion of the training set, which somewhat mitigates the intrinsic sample variability by producing more accurate gradient estimates.

Most deep learning frameworks were designed to train on vast datasets. Since we have many millions of parameters but relatively few samples, augmentation is necessary to artificially inflate the dataset. We capture variation present in anatomy and acquisition by including a mix of rotation (about the axial plane), horizontal flipping and random cropping.

All results are determined by four-fold cross validation on stratified training and testing sets, allowing us to evaluate the network on the entire dataset. The limited size of the dataset introduces an upper bound on overall binary classification accuracy of 92.45% (p-value 0.05), and just 84.8% on a single fold.

5   Results

Metrics were recorded when the loss was lowest for each fold. We will refer to Table 4, containing the results for the combined predictions over the whole dataset, as well as the best performing fold, and detailing the effect of the attention mechanism. It also compares the two levels of localisation that distinguish the *Localised* and *Generic* datasets.

Table 4. Best and average cross-fold binary classification performance for all application variants (formatted by precision/recall, and where A and H denote the abnormal and healthy classes respectively)

Attention		Generic region		Localised region	
		A	H	A	H
✗	Average	0.61/0.20	0.62/0.91	0.76/0.69	0.79/0.85
	Best	0.73/0.47	0.71/0.88	0.93/0.82	0.89/0.96
✓	Average	0.59/0.14	0.61/0.93	**0.79/0.80**	**0.86/0.85**
	Best	0.60/0.35	0.66/0.84	**0.94/0.94**	**0.96/0.96**

In most cases performance is reduced on the underrepresented class of abnormal patients. Moreover, performance is signficantly increased on localised data, and achieved best performance with attention mechanisms - this variant achieves weighted f-1 score **0.83**, demonstrating a strong correlation with the MaRIA score.

However, there is a large disparity between the best fold and the cross-fold average. In fact, the performance of any given fold was found to be highly dependent on the difficulty of the test set. Here we consider the difficulty of the abnormal samples only, assuming that healthy individuals present similar difficulty. We find that difficulty of the best fold was merely 31.3 while the worst was 42.3. Moreover, for the *Localised* variant with attention mechanisms, the average difficulty of incorrectly predicted abnormals was high, at 51.78, and of the seven severely inflamed individuals none were incorrectly classified (see Table 5). Classification power consistently increases with inflammation severity.

The limited size of the dataset introduced severe overfitting in training, forcing us to restrict the depth of the network and degrading overall performance. Furthermore, larger networks performed worse on the *Generic*, or population-specific, variant due to the reduction in SNR. This introduced difficulties in comparing variants on a standardised architecture.

5.1   Attention

Attention mechanisms were found to exacerbate overfitting in scenarios with a low SNR but otherwise boosted performance. This can be seen by observing that attention boosts performance on the *Localised* dataset but degrades it

Table 5. Classification accuracy of best performing variant per inflammation class (for class support refer to Table 2)

Inflammation	Severe	Moderate	Mild	Healthy
Accuracy (%)	100.0	86.2	70.6	85.0

on *Generic*. We theorise that attention mechanisms can only become effective techniques to identify salient, local features within a network if the additional parameters they introduce are not accidentally misused for overfitting. There is evidence for this since the lowest cross entropy achieved by the best performing fold on the *Generic* dataset increased from 0.565 to 0.619 with the addition of attention mechanisms. Referring to Fig. 2, it also assisted us in *debugging* our network by highlighting that zero-padding allows the network to localise to regions that can be overfit on, such as bordering tissue (see Fig. 2a); mirror padding solves this issue. From Fig. 2b we deduce that the attention mechanism successfully identifies the relevant bowel section, reinforcing our confidence in the diagnoses.

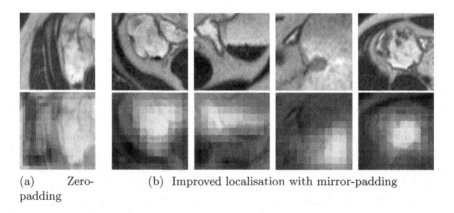

(a)   Zero-padding   (b) Improved localisation with mirror-padding

Fig. 2. Attention maps on the *Localised* dataset (original slice and with attention overlayed on top and bottom rows respectively)

6   Discussion and Conclusion

In this work we demonstrated that a generic deep learning network, trained on a very small MRI dataset, correlates strongly to the MaRIA score, the current clinical standard, while requiring a fraction of the preprocessing by the radiologist. However, the framework is not without limitations in that performance is highly dependent on the level of localisation used in preprocessing and the difficulty rating of the classification at hand. Furthermore, the low dimensionality of the output variable introduces statistical upper bounds on classification power and increases overfit. In this paper the evaluation criteria is solely based on expert radiologist assessment of the MRI data. Validation through colonoscopy is subject to future work, pending ethical approval.

Despite this, we observed very high classification power on the moderate to severely inflamed individuals, suggesting that this algorithm could provide secondary diagnoses to the radiologist in order to flag potentially missed cases. Overall, this pilot study highlights that deep learning is a very promising technique as a method for diagnosing disease in the bowels, and indicates that a larger dataset should continue to be collected for further evaluation. Finally, the limitations encountered through predicting low-dimensionality data might be alleviated by instead automating segmentation of the terminal ileum using deep learning, as a precursor to diagnosis. Thus, we recommend that terminal ileal ground-truth segmentations also be collected.

Acknowledgements. This research was kindly supported by Intel and hardware donations from Nvidia.

References

1. Chang, C.W., Wong, J.M., Tung, C.C., Shih, I.L., Wang, H.Y., Wei, S.C.: Intestinal stricture in Crohn's disease. Intest. Res. **13**(1), 19 (2015). https://doi.org/10.5217/ir.2015.13.1.19
2. Donagh, C., Walshe, T.M., Roche, C., Lohan, D., Cronin, C.G., Murphy, J.: Potential Pitfalls in MRI enterography-a pictorial review. Learning objectives (2012). https://doi.org/10.1594/ecr2012/C-2046, www.myESR.org
3. Fu, Y., et al.: A novel MRI segmentation method using CNN-based correction network for MRI-guided adaptive radiotherapy. Med. Phys. **45**(11), 5129–5137 (2018). https://doi.org/10.1002/mp.13221
4. He, K., Zhang, X., Ren, S., Sun, J.: Deep residual learning for image recognition. In: Proceedings of the IEEE Conference on Computer Vision and Pattern Recognition, pp. 770–778 (2016)
5. Litjens, G., et al.: A survey on deep learning in medical image analysis. Med. Image Anal. **42**(December 2012), 60–88 (2017). https://doi.org/10.1016/j.media.2017.07.005
6. Lundervold, A.S., Lundervold, A.: An overview of deep learning in medical imaging focusing on MRI. Zeitschrift für Medizinische Physik **29**(2), 102–127 (2019)
7. Mahapatra, D., et al.: Automatic detection and segmentation of Crohn's disease tissues from abdominal MRI. IEEE Trans. Med. Imaging **32**(12), 2332–2347 (2013). https://doi.org/10.1109/TMI.2013.2282124
8. Rimola, J., et al.: Magnetic resonance for assessment of disease activity and severity in ileocolonic Crohn's disease. Gut **58**(8), 1113–1120 (2009). https://doi.org/10.1136/gut.2008.167957
9. Schlemper, J., et al.: Attention gated networks: learning to leverage salient region-sin medical images. Med. Image Anal. **53**, 197–207 (2019). https://doi.org/10.1016/j.media.2019.01.012. http://www.sciencedirect.com/science/article/pii/S1361841518306133

Support Vector Based Autoregressive Mixed Models of Longitudinal Brain Changes and Corresponding Genetics in Alzheimer's Disease

Qifan Yang, Sophia I. Thomopoulos, Linda Ding, Wesley Surento,
Paul M. Thompson, Neda Jahanshad[✉], and for the Alzheimer's
Disease Neuroimaging Initiative

Imaging Genetics Center, University of Southern California,
Marina Del Rey, CA 90292, USA
{qifan.yang, sthomopo, lindadin, wsurento, pthomp,
neda.jahanshad}@usc.edu

Abstract. Longitudinal data used as repeat measures may capture the proportion of total variance due to genetic factors with greater sensitivity. However, for brain imaging in studies of older adults, there is a steady decline of brain tissue. It is important to establish such estimation methods using longitudinal data, while properly modeling within-subject variation and rate of tissue atrophy. However, to date, neuroimaging studies have been limited to using only two timepoints, and have not considered diagnostic-specific trends in clinically heterogeneous samples. Modeling temporal patterns of brain structure specific to neurodegenerative disease, while simultaneously assessing the contribution of genetic and environmental risk factors, is essential to understanding and predicting disease progression. We use data from the Alzheimer's Disease Neuroimaging Initiative (ADNI) to model the genetic effects on brain cortical measurements from three repeated measures across two years. We refine our model for specific diagnostic groups, including cognitively normal elderly individuals, individuals with mild cognitive impairment and AD, and then distinguish between those who remain stable or convert to AD. We propose a support vector based, longitudinal autoregressive linear mixed model (ARLMM) for long-term repeated measurements, offering greater sensitivity than cross-sectional analyses in baseline scans alone.

Keywords: Longitudinal mixed model · Alzheimer's Disease · Support vector machine

1 Introduction

Compared to the standard cross-sectional approach for neuroimaging studies, longitudinal clinical designs give important insights into temporal dynamics of the normal aging process or an underlying disease progression, and provide significantly increased statistical power by reducing confounding effects of within-subject variability [1–3]. Once a complex trait is established as heritable [4], identifying specific genetic markers

© Springer Nature Switzerland AG 2019
I. Rekik et al. (Eds.): PRIME 2019, LNCS 11843, pp. 160–167, 2019.
https://doi.org/10.1007/978-3-030-32281-6_17

that explain trait variability may help researchers gain deeper insights into its underlying biological mechanism and offer new targets for therapies.

Most disorders of the brain are still without cures and many lacking effective treatments and there is an increasing prevalence of later life cognitive impairment and neurodegenerative disorders. There is, therefore, an urgent need to understand the genetic and environmental risk factors that may attribute to abnormal brain decline. Longitudinal brain imaging studies such as the multi-site North American Alzheimer's Disease Neuroimaging Initiative (ADNI) [5] acquire clinical, imaging, genetic and biospecimen data that help track cognitive change over time, and aim to identify brain imaging biomarkers that will help predict such changes in clinical populations.

In genome-wide association studies (GWAS) of human disorders, including Alzheimer's disease (AD) [6], an association across each of the millions of single nucleotide polymorphisms (SNPs) across the genome is conducted to identify specific genetic loci that may confer increased risk for the disorder. Individually, SNPs typically explain less than 1% of the population variance in a trait [7], so GWAS usually involve tens to hundreds of thousands of individuals. Polygenic risk scores (PRS), with weights derived from the results of a large and well-powered GWAS of individuals (as training data), can be used to provide an aggregate measure of an individual's genetic risk for individuals from an independent (testing) dataset [8]. Other than large-scale biobanks that include brain imaging [9], single cohort studies of under ten thousand subjects are typically not well powered for performing GWAS. Therefore, instead of single-SNP analyses, here we assess the effect of PRS for AD on brain measures.

In most longitudinal imaging studies of brain aging and neurodegeneration, including most other ADNI publications, researchers only use two timepoints – the baseline and a single follow-up – to assess the longitudinal changes as a simple difference. As there are relatively small sample sizes in neuroimaging studies compared to other epidemiological or genetic studies, this can unnecessarily reduce the available power. Repeat measures taken on a single individual at different times are almost always correlated, with measures taken closer in time being more highly correlated than measures taken further apart in time [10]. Ignoring the dependence of measures both within and across individuals may increase Type 1 error and reduce statistical power [11], especially when there is a known correlation structure such as the genetic relationship matrix (GRM) [4], which captures the genetic similarity across individuals using genome-wide polymorphism information, or other mixed confounding effects such as MRI scanner differences. Furthermore, in longitudinal studies of multi-diagnostic populations, such as ADNI, an additional correlation structure may exist between the time-series of observations from individuals within the same diagnostic classification. Accounting for this diagnostic-specific correlation structure may help to map the trajectories of decline with greater specificity, and improve precision in statistical inferences of genetic or fixed-effects risk variables [10]. This is particularly important for those individuals who ultimately convert to Alzheimer's disease, allowing us to better identify at-risk individuals before the onset of dementia.

Mixed models that are appropriate for longitudinal genetic applications, such as GWAS, include longitudinal MMHE [3] and PMALT [11]. Longitudinal MMHE is based on moment matching learning. This method impressively speeds up the training process, but heavily relies on the large data size to ensure accuracy. PMALT is a prospective likelihood score test based on mixed models, which introduces a random effect with an exponential covariance structure to model the phenotypic autocorrelation, while does not focus on temporal structure modelling specific to different clinical settings.

To the best of our knowledge, genetic influences on brain structure have not been modeled with more than two timepoints, nor have prior efforts attempted to ensure proper modeling of known genetic, environmental, and diagnostic sources of correlation across individual measurements. To better understand disease progression trajectories by modeling spatial-temporal brain structure patterns, we propose a support vector based, longitudinal autoregressive linear mixed model (ARLMM) for long-term repeated measurements. This model considers an autoregressive random effect to account for diagnostic and site variabilities, and in modeling effects of PRS on cortical thickness in ADNI. We also analyzed the genetic associations separately in individual diagnostic groups, as the ε-insensitive loss, implemented in support vector regression (SVR) [12] is applied for smaller samples.

2 Support Vector Autoregressive Mixed Model

2.1 Model Specification

We consider the matrix form of the Autoregressive Linear Mixed Model (ARLMM)

$$Y = Xb + g + t + s + e \tag{1}$$

$$g \sim N\left(0, \sigma_g^2 \Sigma_g\right), t \sim N\left(0, \sigma_t^2 \Sigma_t\right), s \sim N\left(0, \sigma_s^2 \Sigma_s\right), e \sim N\left(0, \sigma_e^2 I\right) \tag{2}$$

Suppose we have imaging phenotypes, covariates and genotype data for N individuals, each with n ($n \geq 3$) repeated measurements. Y (size $nN \times 1$) denotes imaging phenotypes chronically, and X (size $nN \times q$) denotes a covariate matrix, which may include time-varying covariates such as age, and static covariances such as sex [11]. When testing PRS association, we represent the PRS as a vector x (size $nN \times 1$) as follows, while β (constant) and b (size $q \times 1$) are coefficients

$$Y = Xb + x\beta + g + t + s + e \tag{3}$$

Brain structural measures are denoted as Y, which is modeled as the sum of the linear trend and Gaussian-distributed random effects. The distributions of random effects are specified by unknown variance components $\sigma_g^2, \sigma_t^2, \sigma_s^2, \sigma_e^2$; and through known or modeled relationships including the GRM Σ_g, block-diagonal matrices Σ_s and Σ_t, and the identity matrix I. There are two types of random effects: within-subject variations g and t, and between-subject s and e. g and t represent genetic relatedness and autocorrelation for one single individual's phenotypes over time, respectively; s corresponds to the between-subject variations due to scanner differences, where each subblock of Σ_s is a matrix with every entry of 1's, corresponding to individuals whose images were acquired on the same scanner. e represents between-subject environmental errors.

There are 5 diagnostic categories for the ADNI participants across the three assessed time points, corresponding to stable clinical diagnoses and those who convert: (1) stable cognitively normal controls (sCN); (2) stable mild cognitive impairment (sMCI); (3) early converting MCI (ecMCI): those who were categorized as MCI at baseline, but converted to AD by the time of their first follow up at 12-months and remained AD at 24-months; (4) late converting MCI (lcMCI): those who were categorized as MCI but converted to AD between the 12 and 24-month follow-ups; and (5) stable AD (sAD).

To account for the diagnosis-specific time-varying correlation, we assume Σ_t contains unknown parameters. We hypothesize that for subject i at time-point j, t_{ij} is proportional to the most recent disease progression information t_{ij-1} under a first-order autoregressive model, for which autoregressive fluctuations are absorbed by the between-subject environmental errors.

The disease progression parameter, ρ_{ij} is assumed to be only dependent on the diagnostic classification at time point j and $j-1$. Thus, it is specified for each diagnostic group: the stable groups α (CN -> CN), β (MCI -> MCI), γ (AD -> AD), and conversion θ (MCI -> AD). For each subject, the block-diagonal submatrices of Σ_t have the following form:

$$\begin{bmatrix} 1 & \alpha & \alpha^2 \\ \alpha & 1 & \alpha \\ \alpha^2 & \alpha & 1 \end{bmatrix}, \begin{bmatrix} 1 & \beta & \beta^2 \\ \beta & 1 & \beta \\ \beta^2 & \beta & 1 \end{bmatrix}, \begin{bmatrix} 1 & \gamma & \gamma^2 \\ \gamma & 1 & \gamma \\ \gamma^2 & \gamma & 1 \end{bmatrix}, \begin{bmatrix} 1 & \beta & \beta\theta \\ \beta & 1 & \beta \\ \beta\theta & \beta & 1 \end{bmatrix}, \begin{bmatrix} 1 & \theta & \gamma\theta \\ \theta & 1 & \theta \\ \gamma\theta & \theta & 1 \end{bmatrix} \quad (4)$$

2.2 Parameter Estimation

Step 1. To obtain robust and efficient estimates of component variances $\sigma_g^2, \sigma_t^2, \sigma_s^2, \sigma_e^2$, we apply a L2-regularized squared ε-insensitive loss on projected phenotypic covariance and projected random effect covariances. Here, P is an orthogonal projection matrix which satisfies $P^T[X, x] = 0$, and projects out both covariates and PRS in Eq. (3). Component variance estimates can be obtained by minimizing L2-regularized L, via a stochastic gradient descent:

$$L = max\left(0, \left\| P^T YY^T P - \sigma_g^2 P^T \sum_g P - \sigma_t^2 P^T \sum_t P - \sigma_s^2 P^T \sum_s P - \sigma_e^2 I \right\|_2\right) \quad (5)$$

In our implementation, we solved the optimization problem in stable clinical groups first, then infer parameter θ in the converter groups with estimated disease progression parameters α, β and γ.

Our optimization strategy may be viewed as one of the moment matching methods of mixed models [3], but simultaneously learns the unknown component variance and disease progression parameters, which may provide a better description of the disease progression trajectories for stable and converting groups. Furthermore, squared-ϵ insensitive loss ignores any training data close to (within a threshold ϵ) the predicted phenotypic covariances, allowing slightly misspecifying covariance structures of random effects, which may reduce the bias of fixed in both small and large samples [10]. Instead of minimizing the observed training errors, squared-ϵ insensitive loss attempts to minimize the generalization error bound [12], an important consideration for reproducibility in datasets of clinical populations.

Step 2. After the covariance structures for all random effects are determined, the coefficient b and β (only in the PRS association) has a closed form, which minimizes the L2-loss. Delete-one jackknife resampling provided estimate standard errors and p-values for all associations.

3 Experiments

3.1 Alzheimer's Disease Neuroimaging Initiative (ADNI)

T1-weighted MRI brain imaging data from 593 ADNI participants scanned at baseline, 12 months and 24 months was used in this analysis. Images were acquired on 1.5T and 3T scanners during ADNI1 and ADNI2 respectively, using similar imaging parameters [5]. We excluded subjects with non-matched or missing genetic information and inaccurate FreeSurfer measures. If additional follow-up information was available, any individual who converted (from CN to MCI or AD and MCI to AD) after the 24-month follow-up was excluded. The demographic information for the diagnostic groups is detailed in Table 1.

We extracted cortical thickness measures from 34 target regions of interest (ROIs). As the spatial-temporal pattern of brain atrophy in ADNI is complex and highly variable, longitudinal cortical thickness measures were processed with a density based spatial-temporal clustering algorithm, DBSCAN, to remove outliers [13].

3.2 Polygenic Risk Scores and Genetic Relationship Matrix

To estimate pairwise genetic similarity in ADNI, the GRM was built with 644,855 SNPs (minor allele frequency >0.01) [4]. Weights for PRS were determined from the stage 1 results of the International Genomics of Alzheimer's Project (IGAP) GWAS [14], which were considered as training set to derive the SNP effect size. To assess only the impact of the most significant SNPs, we used SNPs that met a p-value threshold of 0.00001 to calculate PRS from the most significant SNPs in the GWAS. It is important to mention the ADNI participants used in this analysis were not included in the stage 1 of the IGAP GWAS from which weights were derived.

3.3 Experimental Settings

Our ARLMM was applied for each of the five clinical groups to determine any association between PRS and average cortical thickness for left and right average cortical thickness from 34 FreeSurfer defined ROIs; covariates included age, sex, ICV, and a dummy variable for field strength. We compared the proposed longitudinal mixed model with its two close cross-sectional counterparts including multivariate linear regression (LR) and support vector regression (SVR) using only data collected at baseline.

To prevent overfitting and improve the model's predictive performance, we used a 10-fold cross validation by computing the mean squared error (MSE), with a validation-based (10%) early stopping.

Table 1. Demographic information for each diagnostic group assessed after removing outliers.

Diagnosis trajectory	Sample size	Sex (F/M)	Age
Stable control: sCN (CN-> CN-> CN)	199	97/102	75.48 ± 5.44
Stable mild cognitive impairment: sMCI (MCI-> MCI -> MCI)	173	68/105	73.20 ± 7.22
Stable AD: sAD (AD-> AD-> AD)	99	45/54	75.28 ± 7.69
MCI converting to AD early: ecMCI (MCI-> AD-> AD)	52	22/30	73.62 ± 6.52
MCI converting to AD later: lcMCI (MCI-> MCI-> AD)	70	26/44	74.41 ± 7.99

3.4 Experimental Results

After Bonferroni correction across the 34 cortical regions assessed, PRS was negatively associated with parahippocampal thickness ($p = 0.0007$) in the sMCI group, and positively associated with the thickness of the lingual gyrus ($p = 0.0014$) for the later converting lcMCI (see Fig. 1). Using our approach, disease progression parameters α, β, γ, and θ are significant (Bonferroni corrected) across almost all 34 regions, which may capture the temporal pattern of brain cortical changes in different clinical groups. θ corresponding to the conversion groups is observed to be significantly lower than disease progression parameters α, β, γ for stable groups.

Neither linear regression nor support vector regression detected significant associations between PRS and cortical thickness (all $p > 0.05$). These methods do not capture the temporal correlation in brain structure, and can only be used when there is no relatedness between individuals; we compare our method with these approaches as these are the methods most commonly used when assessing the genetic influences of brain structure.

4 Discussion

Here, we proposed a longitudinal autoregressive linear mixed model suitable to detect genetic associations from clinical subsets of data with three time points. ARLMM could be further extended to allow unequal measurements, as long as there exists one individual measured at no less than 3 time points. A limitation of this method may be that it cannot handle missing genotype or phenotype data, or participants with fewer measurements, which is part of our future directions planned with this model. We also note that the genetic relationship among ADNI participants is not very strong, and therefore this model may show greater promise in cohorts with stronger genetic relationships. Furthermore, to establish the robustness of our model, we can iteratively remove individuals to obtain estimates on minimal bounds of our method, and determine the robustness when only using two time points. In settings where clinical diagnoses of MCI or dementia may be uncertain or not evaluated, subjects may be grouped based on biological signatures, such as amyloid or tau positivity, sex, or environmental exposures, such as pollutant exposures, or education level, if the variables are believed to show different rates of brain atrophy.

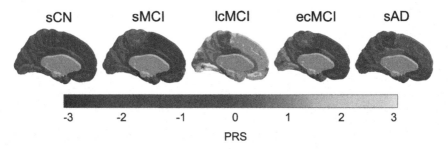

Fig. 1. PRS z-scores of cortical thickness in 34 ROIs for 5 clinical groups are mapped onto the brain. Brain regions where PRS passes Bonferroni corrected threshold of $p < 0.05/34$ are marked with red asterisks. Using ARLMM, PRS is found to be negatively associated with the thicknesses of the parahippocampal cortex ($p = 0.0007$) in the sMCI group, and also positively with the lingual gyrus ($p = 0.0014$) in the later-converting MCI group (Color figure online).

Acknowledgements. We acknowledge support from NIH grant R01AG059874 High resolution mapping of the genetic risk for disease in the human brain. Data used in preparing this paper were obtained from the Alzheimer's Disease Neuroimaging Initiative (ADNI) [5] dataset, which involves both phase 1 and phase 2. Many investigators within ADNI contributed to the design and implementation of ADNI, and/or provided data but did not participate in analysis or writing of this paper.

References

1. Bernal-Rusiel, J.L., et al.: Statistical analysis of longitudinal neuroimage data with linear mixed effects models. Neuroimage **66**, 249–260 (2013)
2. Guillaume, B., et al.: Fast and accurate modelling of longitudinal and repeated measures neuroimaging data. Neuroimage **94**, 287–302 (2014)
3. Ge, T., et al.: Heritability analysis with repeat measurements and its application to resting-state functional connectivity. Proc. Natl. Acad. Sci. **114**(21), 5521–5526 (2017)
4. Yang, J., et al.: Common SNPs explain a large proportion of the heritability for human height. Nat. Genet. **42**(7), 565 (2010)

5. Jack Jr., C.R., et al.: The Alzheimer's disease neuroimaging initiative (ADNI): MRI methods. J. Magn. Reson. Imaging **27**(4), 685–691 (2008)
6. Lambert, J.C., et al.: Meta-analysis of 74,046 individuals identifies 11 new susceptibility loci for Alzheimer's disease. Nat. Genet. **45**(12), 1452 (2013)
7. Munafò, M.R., et al.: A manifesto for reproducible science. Nature Human Behavior **1**, 0021 (2017)
8. Khera, A.V., et al.: Genome-wide polygenic scores for common diseases identify individuals with risk equivalent to monogenic mutations. Nat. Genet. **50**(9), 1219 (2018)
9. Elliott, L.T., et al.: Genome-wide association studies of brain imaging phenotypes in UK Biobank. Nature **562**(7726), 210 (2018)
10. Littell, R.C., et al.: Modelling covariance structure in the analysis of repeated measures data. Stat. Med. **19**(13), 1793–1819 (2000)
11. Wu, X., et al.: L-GATOR: genetic association testing for a longitudinally measured quantitative trait in samples with related individuals. Am. J. Hum. Genet. **102**(4), 574–591 (2018)
12. Basak, D., et al.: Support vector regression. Neural Inf. Process.-Lett. Rev. **11**(10), 203–224 (2007)
13. Schubert, E., et al.: DBSCAN revisited, revisited: why and how you should (still) use DBSCAN. ACM Trans. Database Syst. (TODS) **42**(3), 19 (2017)
14. Ding, L., et al.: Voxelwise meta-analysis of brain structural associations with genome-wide polygenic risk for Alzheimer's disease. In: 14th International Symposium on Medical Information Processing and Analysis, vol. 10975. International Society for Optics and Photonics (2018)

Treatment Response Prediction of Hepatocellular Carcinoma Patients from Abdominal CT Images with Deep Convolutional Neural Networks

Hansang Lee[1], Helen Hong[2(✉)], Jinsil Seong[3], Jin Sung Kim[3],
and Junmo Kim[1]

[1] School of Electrical Engineering,
Korea Advanced Institute of Science and Technology,
Daejeon, Republic of Korea
hansanglee@kaist.ac.kr
[2] Department of Software Convergence,
Seoul Women's University, Seoul, Republic of Korea
hlhong@swu.ac.kr
[3] Department of Radiation Oncology, Yonsei Cancer Center,
Yonsei University College of Medicine, Seoul, Republic of Korea

Abstract. Prediction of treatment responses of hepatocellular carcinoma (HCC) patients, such as local control (LC) and overall survival (OS), from CT images, has been of importance for treatment planning of radiotherapy for HCC. In this paper, we propose a deep learning method to predict LC and OS responses of HCC from abdominal CT images. To improve the prediction efficiency, we constructed a prediction model that learns both the intratumoral information and contextual information between the tumor and the liver. In our model, two convolutional neural networks (CNNs) are trained on each of the tumor image patch and the context image patch, and the features extracted from these two CNNs are combined to train a random forest classifier for predicting the LC and OS responses. In the experiments, we observed that (1) the CNN outperformed the conventional hand-crafted radiomic feature approaches for both the LC and OS prediction tasks, and (2) the contextual information is useful not only individually, but also in combination with the conventional intratumoral information in the proposed model.

Keywords: Computed tomography · Hepatocellular carcinoma · Prediction model · Treatment response · Deep learning

1 Introduction

Hepatocellular carcinoma (HCC) is the sixth most common cancer and the third most common cause of cancer-related mortality worldwide [1]. Radiation therapy is one of the major options for the treatment of HCC, and the task of predicting

© Springer Nature Switzerland AG 2019
I. Rekik et al. (Eds.): PRIME 2019, LNCS 11843, pp. 168–176, 2019.
https://doi.org/10.1007/978-3-030-32281-6_18

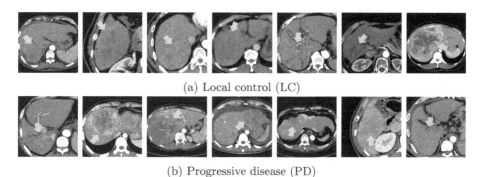

(a) Local control (LC)

(b) Progressive disease (PD)

Fig. 1. Examples of CT images of liver tumors for different LC classes.

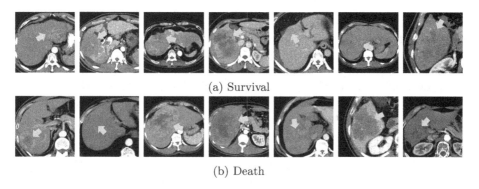

(a) Survival

(b) Death

Fig. 2. Examples of CT images of liver tumors for different OS classes.

the responses of individual patients on radiation therapy is an important process for establishing a treatment plan [2]. The treatment responses include (1) the local control (LC) which determines whether the tumor size is shrinking after the treatment, and (2) the overall survival (OS) which determines the 24-month survival of the patients after the treatment. In terms of binary classification, the LC prediction can be defined as a task of classifying the patient image into one of {LC,PD} classes, where the LC cases show a decrease or increase within 25% in the tumor size, while the progressive disease (PD) cases show tumor size increase of more than 25%. The OS prediction can also be defined as a binary classification of the patient's survival at a particular period in time, such as 12, 24, or 60 months, after the therapy. Prediction of these types of treatment responses from the patient images before the therapy has been of interest. However, this prediction task is known to be a very challenging problem due to (1) the imaging differences between patients with different responses are ambiguous, and (2) the imaging consistency within the patients with same responses is not high enough to cluster, as shown in Figs. 1 and 2.

A number of works have been suggested to predict the treatment responses of HCC from radiomic feature analysis on the patient images. Tamandl *et al.* [3]

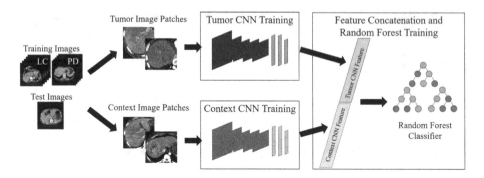

Fig. 3. A pipeline of the proposed method.

analyzed qualitative measures of CT perfusion, e.g. blood flow, blood volume, and portal liver perfusion, for predicting the early response of hepatocellular carcinoma (HCC). Cozzi *et al.* [4] performed univariate and multivariate analysis on CT radiomic features including histogram and texture to predict the LC status and overall survival of HCC patients. Zhou *et al.* [5] predicted the early recurrence of HCC using a logistic regression model with the CT radiomic features of histogram and texture. Shan *et al.* [6] constructed a prediction model for early recurrence of HCC using CT radiomic features of histogram and wavelet transform. These previous studies have shown that (1) the intratumoral texture is especially useful for predicting the cancer treatment responses, (2) the peritumoral radiomic features extracted from the tumor boundary region showed better prediction performance than the conventional intratumoral features and (3) in addition to the intratumoral information, the contextual information of the region where the tumor interacts with the surrounding organs, provides important information for tumor characteristics. However, most of the previous works have constructed the prediction models with qualitative measures and hand-crafted radiomic features, and deep learning has not yet been applied to predict the treatment response of HCC patients.

In this paper, we propose a deep learning method for predicting the LC and OS treatment responses of HCC patients from abdominal CT images. To reflect both intratumoral and contextual information, two CNN models are trained in parallel on the tumor image patches and the context image patches, respectively. The features extracted from the two trained CNNs are then concatenated to be used for training a random forest classifier to predict the local control response of HCC patients.

2   Methods

Our method consists of three major steps: (1) generation of tumor image patch and context image patch, (2) training of CNNs on two types of patches in parallel, and (3) feature combination and training of random forest classifier. A pipeline of our method is summarized in Fig. 3.

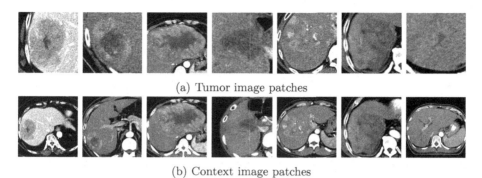

(a) Tumor image patches

(b) Context image patches

Fig. 4. Examples of tumor image patches (top) and their corresponding context image patches (bottom).

2.1   Generation of Tumor- and Context Image Patches

In CNN-based tumor classification methods, the *patch image* which is cropped around the tumor has been used for the CNN input instead of an entire image, to mainly reflecting the tumor information. In this patch generation, the information trade-off occurs depending on the ratio of tumor size to the patch size. If most of the patch region is a tumor, the intratumoral texture information can be mainly reflected, but the context information between the tumor and the organ can be ignored. In contrast, if the patch includes not only the tumor but also some part of the organs, the context information would be strengthened, but the reflection of the intratumoral information would be weakened. Thus, we propose a framework to utilize both information by training both types of patches and combining them.

First, the *tumor image patch* is generated as a squared patch circumscribing the tumor as the conventional image patch for the tumor classifying CNN. The sizes of the tumor image patches with the different sized tumors are resized to match the input size of CNN, i.e. 227×227 pixels so that the CNN training can be tumor size invariant. The generated tumor image patch can reflect the intratumoral texture and tumor shape information to the CNN training.

Second, the *context image patch* is generated as a square patch containing both the tumor and the liver. The generated context image patch can reflect the context information between the tumor and the organ, such as the relative location of the tumor in the liver and the organ information itself. Figure 4 shows the examples of the tumor image patches and their corresponding context image patches.

2.2   CNN Training

With the two types of input image patches, we train a CNN on each of these patches to classify the treatment response classes. The AlexNet [7] is used for training, which is one the most used thin-structured CNNs. The AlexNet consists of five convolutional layers, with the first, second, and fifth convolutional layers followed by max-pooling layers. These convolutional parts are then followed by three fully-connected layers, which finally generate the class label probability from the final layer. In our method, the model is pre-trained on the ImageNet database and is fine-tuned on the given dataset. The transfer learning is performed with the learning rate of $1e^{-4}$, mini-batch size of 20, and maximum epoch of 10.

The trained CNNs can act differently according to the types of input patches. The *tumor CNN* trained with the tumor image patch mainly learns the correlation between the intratumoral texture information and the treatment response, while the *context CNN* trained with the context image patch mainly learns the relationship between the context information and the treatment response.

2.3 Feature Combination and Random Forest Training

As a final step, we combine the information of the tumor CNN and the context CNN and perform classifier training on the combined feature to predict the treatment response. In feature combination, we first extract the two 4096-dimensional features from the fc7 layer of both the tumor and the context CNNs. We concatenate the two features to generate a 8192-dimensional combined feature, which contains both intratumoral texture and context information of the tumor.

In the classifier training, the new classifier is trained on the combined features to classify the treatment response classes. In the proposed method, a random forest classifier is used considering the size of a given dataset. In experiments, the number of trees to construct the random forest was experimentally determined to be 100. The data augmentation is also performed on the training set to generate 5,000 images for each class by random rotation, scaling, and translation.

3 Experiments

3.1 Dataset and Experimental Settings

Our study was approved by the local institutional review board (IRB No. 4-2018-0882). Our dataset includes 171 HCC patients with LC and OS treatment response recordings. In terms of LC, the number of patients diagnosed with *LC* and *PD* in the dataset was 129 and 42, respectively. In terms of OS, the number of patients diagnosed with *Survival* and *Death* in the dataset was 23 and 148, respectively. We used portal phase CT images with the resolution of 512×512, pixels sizes between 0.5×0.5 mm^2 and 0.8×0.8 mm^2, and slice thicknesses between 3 to 5 mm. From CT scans, all tumors and livers were manually segmented by the clinical expert. For each task, the five-fold cross validation was performed on training and evaluating the classifiers.

To evaluate the performances, we performed the internal and external validation for the proposed method. Pipelines of the comparative methods are

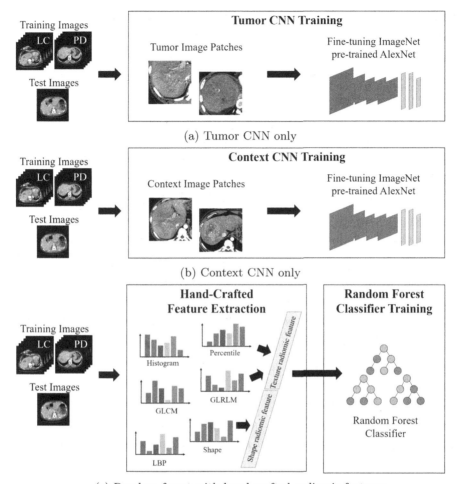

(a) Tumor CNN only

(b) Context CNN only

(c) Random forest with hand-crafted radiomic features

Fig. 5. Pipelines of the comparative methods.

summarized in Fig. 5. As an internal validation, the results of the proposed method were compared with those of (1) the tumor CNN only, and (2) the context CNN only. As an external validation, the results of the proposed method were compared with that of random forest classification with the conventional state-of-the-art hand-crafted radiomic features. In this comparative method, the 67-dimensional hand-crafted radiomic features were used including 7 histogram statistics, 5 histogram percentiles, 14 gray-level co-occurrence matrix (GLCM) features, 22 gray-level run-length matrix (GLRLM) features, and local binary pattern (LBP) features, and 9 shape features. A random forest classifier with the same settings as the proposed method was trained on the hand-crafted features. The detailed list of hand-crafted radiomic features used in the comparison method and the details of the comparative method are given in [8]. The performance was evaluated by comparing the accuracy and the area-under-ROC-curve (AUC) for these comparison methods.

3.2 Results for LC Prediction

Table 1 shows the results of LC prediction for the comparative methods. In internal validation, the context CNN showed slightly lower accuracy than the conventional tumor CNN. This might be since the intratumoral texture and shape are known to be significant criteria for predicting the treatment responses as discussed in the previous studies. However, the proposed method combining the two CNN information improved the accuracy by about 5%p and the AUC by about 2%p compared with the tumor CNN. It can be considered that although the context CNN alone may be less accurate than the tumor CNN, but the context information is complementary to the intratumoral information in prediction so that the proposed method combining both information improved overall performances.

In external validation, the conventional hand-crafted radiomic feature classification showed similar performance as the tumor CNN, while the proposed method outperformed it. It can be considered that deep learning has a high potential in the image-based prediction of treatment response for HCC patients.

3.3 Results for OS Prediction

Table 2 shows the results of OS prediction for the comparative methods. It can be observed that the overall performance of OS prediction is higher than those of LC prediction. It can be considered that the image-based learning provides more information in predicting patient survival than predicting tumor growth.

Table 1. LC prediction results on our dataset.

Methods	Accuracy (%)	AUC
Hand-crafted features	66.08	0.504
Tumor CNN only	67.25	0.567
Context CNN only	66.67	0.572
Proposed	**73.68**	**0.589**

Table 2. OS prediction results on our dataset.

Methods	Accuracy (%)	AUC
Hand-crafted features	81.29	0.611
Tumor CNN only	82.46	0.669
Context CNN only	81.87	0.644
Proposed	**87.13**	**0.706**

In internal validation, the tumor CNN showed slightly higher accuracy and AUC than the context CNN, as shown in the case of LC prediction. The proposed method combining the two CNN information improved the accuracy by 4.67%p and 5.26%p, compared to the tumor CNN and the context CNN, respectively. It can be considered that the intratumoral information and the context information complement each other for the survival prediction as well as the previous LC prediction, so that the proposed method combining both information improved overall performances.

In external validation, the conventional hand-crafted radiomic feature classification showed similar performance to the context CNN, and showed lower performance than the tumor CNN. The proposed method combining two CNN features improved the performance in accuracy by 5.84%p compared to the hand-crafted feature classification. It can be considered that the CNN can provide more useful information than the conventional radiomic hand-crafted features in not only the LC prediction but also the OS prediction task.

In both LC and OS prediction tasks, the values of AUC are overall distributed at low values. It can be analyzed that the models were biased to the larger sized class, i.e. the *LC* class for LC prediction and the *Death* class for OS prediction, due to the class imbalance learning. Further assessment and analysis of the current limitations of the proposed method, and the performance improvement throughout overcoming those problems including the class imbalance remain as future works.

4   Conclusions

In this paper, we proposed a deep learning method to predict LC and OS treatment responses of HCC patients from abdominal CT images. In order to reflect both intratumoral and contextual information, we combined the features extracted from the two CNNs which are trained on the tumor image patch and the context image patch, respectively, and trained a random forest classifier to predict the LC and OS. In experiments, the proposed method achieved higher accuracy compared to not only the existing tumor image-trained CNN, but also the conventional radiomic feature classification. We also verified that the proposed framework can improve the overall performances regardless of prediction target, LC or OS. It can be confirmed that the contextual information is helpful for the prediction of not only the tumor growth but also the survival of HCC patients. Future works include further enhancement of the proposed method throughout the joint learning of CNN and radiomic features and extension of the current classification-based framework to the score regression-level response prediction.

Acknowledgments. This work was supported by Radiation Technology R&D program through the NRF of Korea (NRF-2017M2A2A7A02070427).

References

1. Altekruse, S.F., et al.: Hepatocellular carcinoma incidence, mortality, and survival trends in the United States from 1975 to 2005. J. Clin. Oncol. **27**, 1485–1491 (2009)
2. El-Sera, H.B., Mason, A.C.: Rising incidence of hepatocellular carcinoma in the United States. N. Engl. J. Med. **340**, 745–750 (1999)
3. Tamandl, D., et al.: Early response evaluation using CT-perfusion one day after transarterial chemoembolization for HCC predicts treatment response and long-term disease control. Eur. J. Rad. **90**, 73–80 (2017)
4. Cozzi, L., et al.: Radiomics based analysis to predict local control and survival in hepatocellular carcinoma patients treated with volumetric mudulated arc therapy. BMC Cancer **17**, 829 (2017)
5. Zhou, Y., et al.: CT-based radiomics signature: a potential biomarker for preoperative prediction of early recurrence in hepatocellular carcinoma. Abd. Rad. **42**, 1695–1704 (2017)
6. Shan, Q., et al.: CT-based peritumoral radiomics signatures to predict early recurrence in hepatocellular carcinoma after curative tumor resection or ablation. Cancer Imaging **19**, 11 (2019)
7. Krizhevsky, A., et al.: ImageNet classification with deep convolutional neural networks. In: Proceedings of NIPS, pp. 1097–1105 (2012)
8. Lee, H., et al.: Differentiation of fat-poor angiomyolipoma from clear cell renal cell carcinoma in contrast-enhanced MDCT images using quantitative feature classification. Med. Phys. **44**, 3604–3614 (2017)

Author Index

Printed in the United States
By Bookmasters